模具設計與製造
Mold Design & Manufacturing

五軸銑削數控加工之基礎及實作

Basics and Practices on Five-Axis CNC Milling

王松浩
吳世雄 著

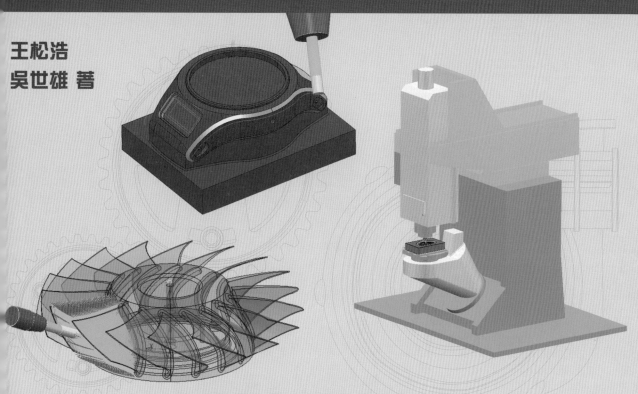

五南圖書出版公司 印行

推薦序

　　機械加工乃是一門基礎且應用性廣之工業技術，該技術之發展層次往往決定了各類產品（例如：機械零件、家用光電產品、儀器設備以及資訊與通訊產品等）之競爭力，該技術層次亦是進入工業大國之一決定性指標。

　　工具機具先進機械加工技術之總整，其市場應用不斷地被開發，從早期機汽車及家電產業，歷增模具、航太與 3C 產業，以及近年新興的生醫、綠能、太空等產業；其技術層次不斷地被提升，今日工具機核心技術之研究，已不再僅侷限於以機構、電控為主體之領域，舉凡資通訊技術（ICT）、光學雷射、影像處理、遠端監控，乃至雲端運算（Cloud Computing）等，皆屬工具機可整合之技術範疇。

　　台灣工具機產值與出口值，歷年皆排名世界前五，超越美國、瑞士以及鄰近的韓國，唯台灣工具機產業結構，目前仍遭遇基礎技術不足、通路受制於人、產品同質性高以及匯率與 FTA 問題等問題。台灣工具機平均單價約為德國及瑞士的 1/2，日本的 3/5，在國際市場認知上，向以中低級產品為主，如何提升品級乃是當前之最大挑戰。此外，從 2013 年第一季開始，台灣工具機出口成長幅度受到日幣大幅貶值的影響，與日本價差從原先 25~30% 縮減為一成多，使得輸出歐美中高階產品之工具機業者首當其衝，亟需基礎技術之廣度教育與提升，再造榮景。

　　本書《五軸銑削數控加工之基礎及實作》是一本介紹多軸加工技術之專業書籍，個人才疏學淺，得能為其寫序實感榮幸，尤其本書之出書動機乃為提升台灣工具機基礎技術之教育做貢獻，因此為序時內心倍感高興。本書介紹之範圍涵蓋多軸座標系之學理介紹以及實務加工練習，文體編撰通俗易懂，且佐以中英對照說明，書中加工實例均經實機驗證過，讀著若可依照章節順序研讀，應可快速地進入多軸加工之殿堂。

本書兩位作者均是個人的好友，一位服務於學界、一位服務於業界，兩位作者在國內機械加工專業領域上、同享盛名、學識淵博且得有十數年的合作經驗，在本技術傳承上也得有難得的共識。身為好友，深感驕傲，預祝本書暢銷，並廣受專業人士及學子親睞，得為工具機之基礎技術提升做出貢獻。

林毅欽

美國奧克拉荷馬州立大學機械暨航空工程博士

崑山科技大學機械系教授／系主任暨所長

2015 秋

簡介（Abstract）

近年來電腦技術的突飛猛進導致加工技術日新月異，物聯網所引起的「零庫存」概念對製造業革命性衝擊，要求「小批量多樣化」甚至「隨到隨做」。而電腦控制及運算的能力又為以上的需求提供了無限的可能性。因此近年來電腦輔助製造的進步也可以說跳躍性的。

因應時代的變遷以及產業競爭的需求，基於 CNC 四軸／五軸加工的高效率，高品質及零件多樣化，業界和學校近年來引進越來越多的多軸加工中心，因此現在非常需要有一些相關的比較基礎和實用的教材。此外目前在技術考試證照中加工領域只有丙級和乙級，而還沒有相應的甲級證照，筆者認為一個主要原因是在多軸加工（多軸銑削及車銑複合等）技術的普及／規範化有待於提到一定的高度。因此作者拋磚引玉地編寫這本以實作為導向的「電腦輔助多軸銑削加工」基礎教材。

對於 CNC 四軸及五軸銑削加工，本教材結合目前業界一些常用的電腦輔助加工軟體以及 CNC 控制器，儘量遵循「通俗實用—無師自通」的原則，一步一步地講解從材料及刀具選擇，刀具路徑計算，虛擬加工模擬，虛擬機器模擬，NC 碼輸出，機台工作座標確立，刀具長度測量，直到加工的進行。在電腦輔助加工軟體方面本書透過編程實例來直接導引多軸加工的應用概念，操作案例經由英國 Delcam 公司推出的 PowerMILL 軟體做編寫，它是目前全球市面上廣泛用於 2～5 軸數控銑削加工自動編程系統。該軟體具有易學易用人員培訓快、程式製作時間縮短、機台加工時間縮短、加工表面品質提升、機台／刀具壽命延長與二次開發容易／方便經驗傳承等優勢特點。在加工實作方面，本書則通過常用的日本 Roland 公司 MDX-40A CNC 以及德國 Siemens 公司 SINUMERIK 840D sl 控制介面進行講解。對其它品牌的軟體或控制器介面，讀者應該會很容易的觸類旁通，融會貫通。

此外為適應國際化的大趨勢，本教材儘量以中—英文雙語表達，不僅使外籍生也可以使用，本籍生更可用來提高英語特別是專業英語的水準。

Due to great achievements in computer technology, the requests for "Zero Storage" and "Flexible Manufacturing" gradually become reality. Although more and more multi-axis machine centers have been introduced into companies and schools, there are very few basic textbooks to systematically teach this subject. Therefore, this is our effort to make multi-axis machining process easier to learn.

In this book, major procedures including CAD model import, block material and tool setups, toolpath generation and simulation, NC code exporting and machining simulation are explained step by step. Moreover for real machining, basic machine setup and operation are also illustrated in detail. In this book PowerMill® by DelCAM is used to create toolpaths and G&M codes. To demonstrate practical machining process, MDX-40A® (4-Aixs) from Roland and SINUMERIK 840D sl® (5-Axis) from Siemens are used as demonstrations.

Although the principle of the writing is to make the readers "Learn & Do It Yourself", it would be easier if the reader had some basic ideas about how the 3-Axis CNC milling process works.

In addition, the book is basically written bilingually in Chinese-English, to serve for non-Chinese readers as well as for native-Chinese readers to improve their Professional English.

前言（Preface）

近年來電腦技術的突飛猛進所導致加工技術日新月異，物聯網所引起的「零庫存」概念對製造業革命性衝擊，要求「小批量多樣化」甚至「隨到隨做」。而電腦控制及運算的能力又為以上的需求提供了無限的可能性。因此近年來電腦輔助製造（Computer Aided Manufacturing, CAM）的進步也可以說跳躍性的。Due to great achievements in computer technology, the requests for "Zero Storage" and "Flexible Manufacturing" gradually becomes reality.

因應時代的變遷以及產業競爭的需求，業界和學校近年來引進越來越多的多軸加工中心，因此現在非常需要有一些相關的比較基礎和實用的教材。此外目前在技術考試證照中加工領域只有丙級和乙級，而還沒有相應的甲級證照，筆者認為一個主要原因是在多軸加工（多軸銑削及車銑複合等）技術的普及／規範化有待於提到一定的高度。基於此因以及上司和同事的鼓勵與促使，才拋磚引玉地試圖編寫這本以實作為導向的「電腦輔助性多軸銑削加工」基礎教材。Although more and more multi-axis machine centers have been introduced into companies and schools, there are very few text books to systematically teach fundamentals and practices in this subject. Therefore, this book is an effort to make multi-axis machining process easier to learn.

詳述 CNC 多軸加工操作的主要步驟，Steps for the Multi-Axis Milling explained in the book：

- CAD 模型的輸入，Import and place the CAD model;
- 工件材料及刀具選擇，Establish block material and cutting tools;
- 刀具路徑計算，Create toll paths based on the geometry and cutting strategies;
- 虛擬加工模擬，Virtual machining simulation to confirm the toll paths;
- 虛擬機器模擬，Virtual machine simulation to prevent collision;
- NC 碼輸出，Export NC (G&M) codes;
- 機台工作座標確立，Establish Work Coordinate on real multi-Axis milling machine;
- 刀具長度測量，Measure toll length compensations;

- 四軸或五軸銑削加工的進行，Start to machine.

本教材特點，Special features of this book：

- 本教材力求通俗易懂，試圖使讀者順著書中步驟就能夠進行軟體 / 硬體實作；Easy to read, follow and practice;
- 如果讀者具備一些基本的三軸 CNC 銑床加工知識及經驗，則實作起來會更加得心應手；Feel much easier if the reader has at least basic idea or experience in three-axis milling practice.
- 本教材主要內容及操作步驟盡可能以中—英雙語展現，以利外籍生的教學以及本籍生英語程度的提高。Written bilingually in Chinese-English, this book is to serve non-Chinese readers as well as for native-Chinese readers to improve their Professional English.

　　本書撰寫過程中，得到了各層面的激勵及促使。例如崑山科技大學工學院及機械系長官不懈的鼓勵；達康科技股份有限公司（Delcam Taiwan）的總經理以及西門子科技股份有限公司臺灣分公司（Siemens Taiwan）相關主管的認可。崑山科技大學機械系與達康科技公司（Delcam Taiwan）在 CNC 加工方面密切聯合之研究與教學已有十餘年之久。現在合作編寫此教材，也算是多年以來合作結晶之展現吧。During the preparation and writing of the book, encouragements from the colleagues of Kun Shan University were the source of energy. Moreover, approves from both managements of Delcam Taiwan and Siemense Taiwan were also appreciated. The close cooperation between Kun Shan University and Delcam company dates back over more than ten years, and therefore this book could be a very good summery.

　　此外作者還要特別感謝，Special thanks go to：

- 崑山科技大學蒙古國留學生貝德傑，尤瑞雅和哥斯大黎加留學生馬克斯對實際機台操作步驟以及零件實際加工之驗證。
  Foreign Students Turbat Byambadelger, Uranzaya Janchivdorj and Max Aguilar Camacho in Kun Shan University (Taiwan) for their tireless efforts in editing the writing and confirmation on machine setup procedure as well as real

machining.

- 達康科技股份有限公司邱楷婷（Kai）在本書編寫及程式編制過程中的大力幫助，Mrs. Kai of Delcam Corporation for her great helps in CAM programing and posting out NC codes;
- 新營高工模具科高賦源老師對銑削加工常用 G&M 碼的確認。Mr. F.Y. Gao from National XinYing Industrial Vocational High School, for his confirmation on most used G&M codes in milling operation.

由於時間比較倉促加上又是雙語注釋，在版面及翻譯方面難免有不盡人意之處，還望讀者多多指正。Because of the time limit and the bilingual contents, the book is much less than perfect. Comments and suggestions are warmly welcome to make the second edition better.

王松浩，吳世雄

2015 夏於台南

目　錄

1

緒　論（Introduction）

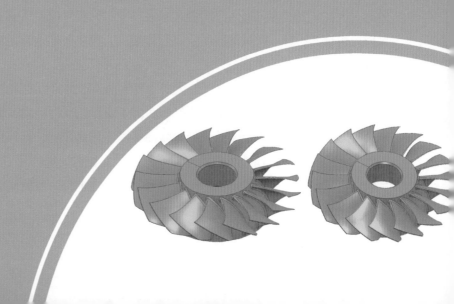

廣義地說機械工業是典型的技術密集工業，也是國家的樞紐工業。其包括：1. 一般機械，例如紡織機械、化工機械、工具機等；2. 電氣機械，用於電力生產及輸配電用之設備，以及資訊與通訊產品等；3. 運輸工具，包括汽車、機車、自行車、火車、船舶、飛機等及其附件；4. 精密機械，包括照相機、望遠鏡、醫療設備、鐘錶、光學儀器、檢驗測試設備等；5. 金屬製品，包括照相機、望遠鏡、醫療設備、鐘錶、光學儀器、檢驗測試設備等。故機械工業有「工業之母」的美譽。Basically speaking, machinery industry is very technology-intensive and also the fundamental industry of many countries. Therefore, it is usually called the "Mother of Industry".

一、機械加工製造業的重要性
（The Importance of Mechanical Manufacturing Industry）

國內通稱之機械工業則僅指一般機械製造業，為狹義的機械工業。係各產業直接於生產之機械設備。範圍包括：工具機、產業機械、通用機械、動力機械及機械零組件等。

由於許多機械設備與下游應用產業相關技術存在著密不可分的關係，因此，下游應用產業對於生產技術之要求，常常成為帶動上游機械工業業著產品改良之主要動力，且機械工業產品品質亦對下游應用產業對外競爭力影響甚巨。機械工業的發展也常被用來衡量一個國家工業進步的程度，為了國家的經濟發展，世界各國無不殫精竭去發展其機械工業。

綜觀機械工業產業特性如下：

1. 為國家工業化程度之指標；
2. 機械工業具有較高的加工層次；
3. 為融合專業科技的工業；
4. 為高度依賴專業人才的產業；
5. 產品週期逐漸縮減；
6. 為技術密集的產業。

全球機械工業主要生產國家以日本、德國、美國為主，而上述國家乃世界上著名之工業大國，因此，台灣也已邁入已開發國家之林，如何提昇國內工業化程度，機械工業之發展占有舉足輕重的影響。機械業發展所需的專業人力較多，亦須有足夠的資金以資配合，是一種技術及資本密集之工業。

台灣在經濟發展的過程中，機械業的發展扮演了相當重要的角色，其主要原因如下述。

1. 台灣的機械工業自 1982 年起，與電子資訊工業同列為策略性工業，就融資、租稅及技術等方面，給予優惠及協助以鼓勵廠商，希望透過政府的輔導加速機械工業開發，以帶動整個工業的發展，進而達成整體產業全面升級之目標。歷經了三、四十年的歷史，發展迄今無論在國內市場或國際市場上已占有舉足較重之地位，且對產業界之發展、整體製造水準之提升，重要性亦逐年加重。

2. 機械業素有「工業之母」之稱，其與下游產業間的互動關係密切，而機械產品品質對產業競爭力影響甚鉅，因此，又成為衡量一國進步與否的指標；如何透過機械品質改良，重劃產業發展藍圖，再藉由產業技術的提升，推動機械的研發，將是產業機械生涯規劃中的重要課題。[1]

二、加工機（工具機）行業的重要性（The Importance of Machine Tool Industry）

加工機（工具機）是用來對工件的外形尺寸或性能進行改變的過程。其主要功能實為製造機器的機器，也因此常被被稱為「工作母機」。按被加工的工件處於的溫度狀態，分為冷加工和熱加工。一般在常溫下加工，並且不引起工件的化學或物相變化，稱冷加工。一般在高於常溫狀態的加工，會引起工件的化學或物相變化，稱熱加工。冷加工按加工方式的差別可分為切削加工和壓力加工。熱加工常見有熱處理、鍛造、鑄造和焊接。

台灣工具機廠商亦是少數以自有品牌行銷國際的重要產業之一。台灣工具機產業在技術整合、產品管理、行銷策略和服務方面具有相當優勢，近年來在政府積極輔導與產業界在提升精密機械之創新研究技術下，工具機產業在高階技術實力不斷精進，且各廠商在國際市場及行銷均有相當豐富的經驗，因此台灣工具機產業具有優良的發展前景。在創新研發的部份，工具機廠商在台灣精品獎、中小企業小巨人獎、工具機研發創新獎 …… 等獎項屢次獲獎，可見台灣工具機廠商持續在產品創新與技術研發仍不斷的進步；工具機廠商藉由積極研究創新，提升產品之品質、設計及形象，並因此提高我國產品附加價值。其中又以上銀科技、高鋒工業、台中精機等在諸多產品創新的獎項屢榮獲殊榮，可見台灣工具機廠商在研發、設計、品質、行銷等均積極投入並有卓越的成果。

在海外布局方面，除了在全球五大洲皆有經銷商外，上銀、台中精機、東台、亞崴、程泰、金豐、亞德客、福裕、協易、龍澤、油機、協鴻、友嘉等分別在中國大陸、德國、美國、日本、瑞士、捷克、法國、以色列亦有海外子公司；因此台灣工具機產業不只深耕國內，並拓展事業版圖至全球各地。

雖然台灣工具機產業榮景可期，但是所面臨的全球競爭也是十分嚴厲的。工具機產品設計層次可依序往上分為：標準化／無差異化；模組化／差異化；直到融入美學與體驗經濟的個性化／體驗化。而台灣工具機約落在設計模組化，與差異化之間；日本為設計差異化，與個性化間；德國則居於最高位置的設計個性化與體驗化間，造就 3 國生產的工具機存在著極大的價格落差。在競爭激烈的大陸市場中，日系機種售價可比台灣貴 1 倍，德國機更是台灣機的 10 倍之多 [2]。

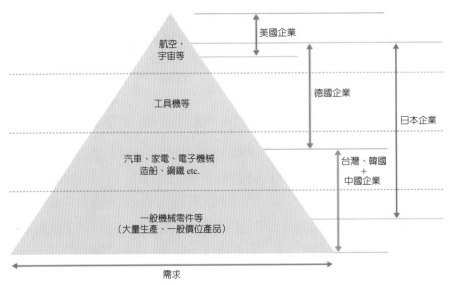

主要生產工具機國家的國際性布局，Major Machine Tool Manufactures[3]

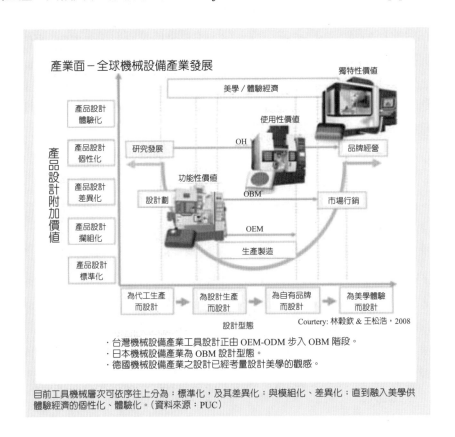

工具機設計的附加價值及價格比較，Price comparisons on Machine Tools [4]

三、多軸加工的重要性（The Importance of Multi-Axis Milling）

導入的思維在於一機多用途的 3 軸高速加工、 3+2 軸定位加工、 5 軸定位鑽孔和 5 軸連續加工等應用優勢，相對於軟硬體技術的提升與成本的降低也是產業界導入的主因考量。

五軸加工機在加工上比較於三軸加工機多了那些優勢？**What are the advantages of 5-Axis over 3-Axis machining**？

· 加工深穴模具，**Deep pocket machining**：允許夾持短刀具加工陡峭側壁或凸島，降低斷刀的風險與減少放電加工。

· 縮短刀具長度，**Shorten necessary tool length**：提高刀具剛性、延長刀具壽命，也提高加工表面精度及品質。

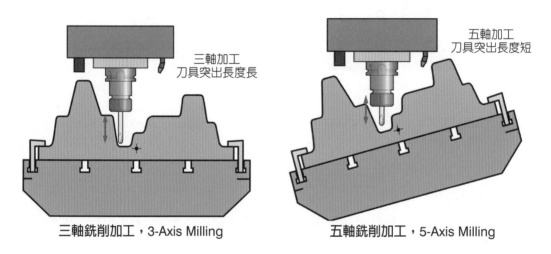

三軸銑削加工，3-Axis Milling　　　五軸銑削加工，5-Axis Milling

· 提高刀具使用率，**Increase efficiency of tools**：透過刀具的刀腹、刀緣做切削並可避開靜點加工。

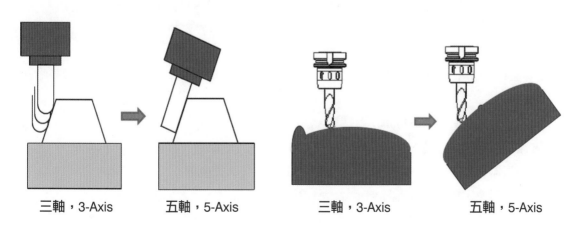

三軸，3-Axis　　　五軸，5-Axis　　　三軸，3-Axis　　　五軸，5-Axis

• 允許加工倒勾區域，**Make it possible to machine overhanging geometries**：減少成形刀的使用，減少後製作流程（EDM 及拋光）。

• 一次工件夾持定位，**Reduce the time and cost for changing fixtures**：利用刀軸的旋轉，完成各部位最適宜的加工方式以減少夾治具的架設及設定。顯著的時間效益在於減少工件重覆裝夾時間，提高了加工精度。

 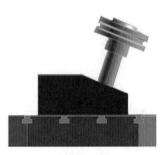

傾斜治具

3 軸加工機　　　　　　　　　5 軸加工機

五軸加工應用基本上有分為 **3** 個種類，**Three major categories for 5-Axis machining**：

• 多面固定軸向加工（**5** 面 & **3+2** 軸），**Multi-Face fixed axis (5-Faces or 3+2) milling**：切削時旋轉軸固定，73% 五軸加工應用採用固定式 3+2 軸加工，廣泛應用於幾何特徵零件加工、車燈模加工及保險桿加工。

• 多軸向鑽孔加工，**Multi-axis drilling**：對於多次翻面的加工孔系，可以一次裝夾就加工完成，如引擎。

• 連續五軸同動加工，**Milling with 5-Axis simultaneous motion**：依加工物的需求，同時轉動兩個旋轉軸及移動三個直線軸進行加工的動作，適合加工葉片 / 深穴模具。

五軸加工機可應用在那些產業，**Applications of 5-Axis machining**：

• 汽車工業，**Automobile**：全尺寸車身模型製作，車身鈑金衝壓模具模面加工，車燈模側花加工，輪胎模具製作。

• 模具業，**Injection Molding**：塑膠模具用於細微清角取代放電，3C 模具，鞋模多軸孔向加工。

• 工具機業，**Machine tools**：導螺杆，滾齒凸輪加工，捨棄式刀具刀把加工。

• 產品開發快速原型，**Rapid prototypes, Mockups**：產品設計打樣，3C，汽機車。

• 醫療器材業，**Medical Device**：齒模，骨釘骨板，人工關節。

• 能源工業，**Energy**：壓縮機葉片，發電機組渦輪扇葉，高效率風扇。

• 造船工業，**Ship building**：高效率船舶推進器漿葉，船身模具。

• 航太工業，**Aerospace**：機身結構框架，單片機翼表面，其它特殊零件加工。

　　五軸加工機及車銑複合加工機是近幾年來機械商研發的重要方向，因應目前加工零件的多樣少量化及複雜零件高精度的加工訴求，針對加工件可一次性的在機臺上完成所有工序，減少因拆裝所造成的誤差，五軸加工機及車銑複合加工機可說是首選。但相對於使用者的部分也需要更多一點的投入，如人員的選用及培訓，加工經驗的傳承，選用適合的加工軟體等等……這些，都是重要功課，複合化加工的時代已來臨！機械商及使用者面臨了更多的挑戰，突破極限為我們共同的目標，大家一起共勉之。

　　隨著工具機的車銑工藝持續演進，五軸加工成為新一代的加工技術，然而產業人才普遍難找。針對這種需求，也許可以設立「車銑複合五軸代工中心」，滿足客戶複雜化的加工工藝與複合化的加工應用技術。

　　因應時代的變遷以及產業競爭的需求，基於 CNC 四軸／五軸加工的高效率，高品質及零件多樣化，業界和學校近年來引進越來越多的多軸加工中心，因此現在非常需要一些相關的比較基礎和實用的教材。此外目前在技術考試證照中加工領域只有丙級和乙級，而還沒有相應的甲級證照，筆者認為一個主要原因是在多軸加工（多軸銑削及車銑複合等）技術的普及／規範化有待於提到一定的高度。因此作者拋磚引玉地編寫這本主要以實作為導向的「電腦輔助多軸銑削加工」基礎教材。Due to great achievements in computer technology, the requests for "Zero Storage" and "Flexible Manufacturing" gradually become reality. Although more and more multi-axis machine centers have been introduced into companies and schools, there are very few basic textbooks to systematically teach this subject. Therefore, this is our effort to make multi-axis machining process easier to learn.

參考文獻（References）

1. 謝明瑞，「台灣機械業的發展」，http://old.npf.org.tw/PUBLICATION/FM/091/FM-R-091-009.htm

2. 馮立誠，陳念舜，資源共用　美學加值少燒錢，機械技術 2012/11 月號，原文網址：http://mag.nownews.com/article.php?mag=11-60-15932#ixzz3XBrK3I39

3. http://mag.nownews.com/article.php?mag=11-60-15932

4. http://luckylong.pixnet.net/blog/post/48382478

2

多軸加工基本運動及座標系統（Basic motion and Coordinate System in Multi-Axis Milling）

✎ 2.1 多軸銑削加工基本概述 （Basics for Multi-Axis Milling）

以銑削加工中心機而言，一般來說基本的三個直線移動軸即為 X 軸，Y 軸及 Z 軸，四軸是由基本的三個直線移動軸再加上一個旋轉軸，而五軸則是再加上二個旋轉軸。旋轉軸定義如下：繞著 X 軸旋轉移動的軸向稱為 A 軸，繞著 Y 軸旋轉移動的軸向稱為 B 軸，繞著 Z 軸旋轉移動的軸向稱為 C 軸（下圖）。一般來講，將多於三軸的加工稱為多軸加工。本教材主要著墨於四軸及五軸銑削加工。Generally in 3-Axis milling process, there are three linear movements along X, Y and Z axis。In 4-axis milling, a rotary motion is added, usually about x axis. In 5-axis milling, two rotary motions are added. The names of the rotary motion are based on the axis it is rotation about: X-A, Y-B and Z-C. Any milling process above 3-axis is called multi-axis milling. This book is concentrated in 4-axis and 5-axis milling.

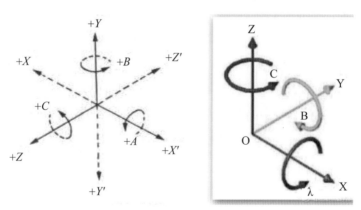

直線移動軸及旋轉軸的定義（Definition of linear and rotary motion in Milling）

一、四軸銑削加工配置（4-Axis Milling Arrangement）

下圖為一般的四軸銑削加工配置，即在 3 軸銑削基礎上加上一個旋轉軸，通常是 A 軸，即 3+1 配置。可以由四軸銑削加工成型非常多形狀的零件，範例如下。The picture blow presents typical 4-axis milling process: a rotary movement is added about x-axis. There are many different shapes can be machined with this process.

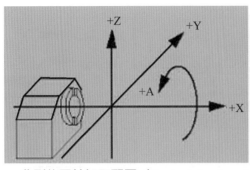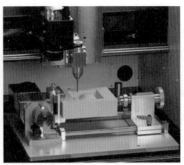

典型的四軸加工配置（Typical 4-axis milling machine arrangement）

Woman body

Chinese guard lion

Queen and knight

Copy of TIICA Medal

KSU towel

Oscar statue

四軸加工應用，崑山科技大學（Work-pieces made with 4-axis milling, KSU）

筆電外殼
製作材質：ABS

玩具槍模型
製作材質：ABS

玩具槍模型打樣
製作材質：ABS 透明採壓克力

腳踏車零件
製作材質：ABS

自知車齒輪
製作材質：鋁

生活日常用品
製作材質：PON

四軸加工應用，達奈美克公司 — 臺灣（Work-pieces made with 4-axis milling）

二、五軸加工配置（5-Axis Milling Arrangement）

五軸加工機的種類有很多，一般常見之正交型（Orthogonal）以及非正交型（Non-Orthogonal）兩大類別的五軸加工機進行分類，正交型是指旋轉軸與線性軸皆正交，非正交型（斜軸型）指旋轉軸與線性軸不一定正交。關係上的因素以下述來進行分類：(1) 旋轉軸是帶動主軸或是帶動工件，(2) 兩個旋轉軸的驅動關係，(3) 旋轉軸的軸向，概述說明如下：There are many types of 5-Axis milling machines on the market, based on the requests of real machining environment. Basically however, there are two categories: Orthogonal and Non-Orthogonal, depending on if the rotary axis is orthogonal to linear axis or not.

Spindle 型（主軸頭擺動型）、Table 型（工作臺旋轉型）及 Hybrid 型（混合型）：五軸加工機的兩個旋轉軸若皆可帶動主軸，則稱為 Spindle 型（主軸頭擺動型），如下圖所示；兩個旋轉軸若皆可帶動工件，則稱為 Table 型（工作臺旋轉型）；若一個旋轉軸帶動主軸，另一個旋轉軸帶動工件，則稱為 Hybrid 型（混合型）。以正交軸主要分類有這三種，若以旋轉軸的軸向關係來細分類正交型五軸加工機將分類成 12 型，在此我們將不深入說明此軸向關係的類型差異。For orthogonal machines, there are three major arrangements:

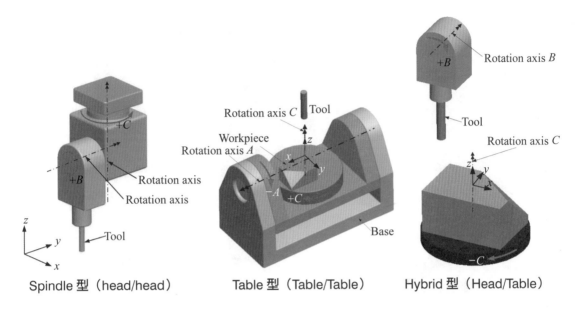

Spindle 型（head/head）　　　　Table 型（Table/Table）　　　　Hybrid 型（Head/Table）

　　非正交型相對於正交型爲複雜，針對雙斜軸大致分成三種，三種型式爲，Compare to regular milling machine arrangements, there Non-Orthogonal axis arrangements. There are basically three types:

　　(1) Spindle 雙斜軸型，(2) Table 雙斜軸型，(3) Hybrid 雙斜軸型。而雙斜軸包含了單斜軸，當旋轉軸向量其中一個分量爲零，則稱爲單斜軸。若是一個單斜軸爲旋轉軸另一旋轉軸平行於線性軸，此機型亦爲市面上可見的機型，以雙斜軸的機型此種機型具有代表性。

雙斜軸型（Inclined Spindles）　　　　雙斜軸型（Tilted Table）

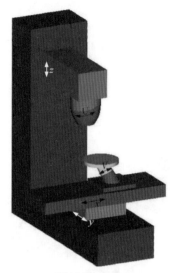
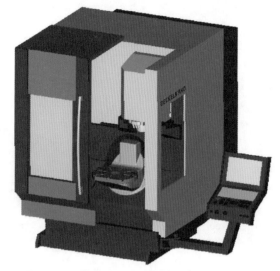

雙斜軸型（Hybrid）　一個旋轉軸為單斜軸另一旋轉軸平行於

線性軸（One Axis inclined and one parallel）

三、五軸加工，五面加工及其他，5-Axis and 5-Face Milling

　　五軸加工機與五面加工機這兩者最大的區別在那？在於是否具有五軸同時同動，The basic difference between 5-Axis and 5-Face milling is if the five axis can move simultaneously or not.

五軸加工機（5-Axis Milling Machine）

　　擁有 3 個線性軸及 2 個旋轉軸，多了兩個自由度來控制刀軸的彈性變化，在 3D 空間的任何位置之曲面或平面均可加工，以 3 個線性軸來決定刀具位置，2 個旋轉軸來決定刀具方向，五軸同時達到指定的位置及方向，無論工件輪廓如何變化，刀具均能保持與工件表面垂直或特定角度。五軸比三軸優勢在於可一次夾持定位，加工複雜的工件：如葉片 impeller...，如下五軸加工應用範例。Because in 5-axis milling, three linear axis determines the position of the tool, two rotary axis determines the orientation of the tool and the five axes moves simultaneously, the tool-axis can be perpendicular to or at any angle to the surface of the workpiece. Not only it can machine regular workpiece without change fixture, it can also machine almost all complex shapes.

五軸加工應用範例（Examples of 5-axis machining）

五面加工機（**5-Face Milling**）

　　在單一次的架設條件下，利用旋轉工作臺或刀具頭作特定角度之定位後再進行二軸或三軸之加工順序。如右圖示對正立方體的五個面進行加工的動作，不具有五軸同時到達定位與方向之功能。In the 5-Face milling process, the tool-axis is aligned to one of the 5 faces try rotating thd tool-axis or the work piece, then stat 3-Axis milling. There is no need of simultaneouse movement with all five axis.

　　除此之外，市面上還有以機械手臂控制進行加工，下圖就是以六軸機械手臂加工的實例，應該是特別適合於大型工件的加工場合。Moreover, robotic-arms are used for large parts. One of the examples of 6-axis robot arm is illustrated in the following picture.

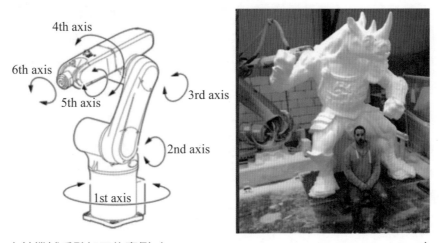

六軸機械手臂加工的實例（An example of 6-axis robot arm used in CNC）

因為本教材討論的範例是以雙旋轉工作臺的立式銑床（Table-Table）為主進行加工，故需要討論一下旋轉工作臺的類型，下圖左為 BA 的旋轉工作臺，B 為繞 y 軸，A 為繞 X 軸的旋轉，主要適用於臥式銑床（Horizontal Machine Center, HMC）。而 AC 則是為立式銑床（Vertical Machine Center, VMC）所設計的。Because Table-Table style is the machine used in this text book, it is necessary to mention the difference between two major dual-rotary-tables. In the following picture, the BA table is designed for horizontal milling center and AC table is for vertical milling center.

BA dual table for HMC

AC dual table for VMC

雙旋轉工作臺的類型（Types of dual rotary table）

2.2 銑削加工之座標系統
（Coordinate system on milling machines）

座標系是用來定義加工機工作平面或工作空間的位置。而位置數據總是預先通過座標系表達的確定點。A coordinate system is used to uniquely identify positions in the working plane or working space of the machine. The position data are always referenced to a predetermined point, which is described by coordinates.

一、基本術語

對每一台工具機，廠商已經為加工機定義了一個永久的參考系 - 即機器座標系。而對每一個工件來說用戶可以自由地為選擇一個座標系：因為控制器知道這兩個座標系之間的關係，因此控制器就能夠正確地將 NC 程式中的位置座標應用於工件的加工。The machine manufacturer specifies a permanent reference system for a machine—the machine coordinate system. The user can freely select a separate coordinate system for each workpiece: the control is aware of the position and origin of this system with respect to the machine coordinate system. This enables the control to correctly apply the position entries from an NC program to the workpiece.

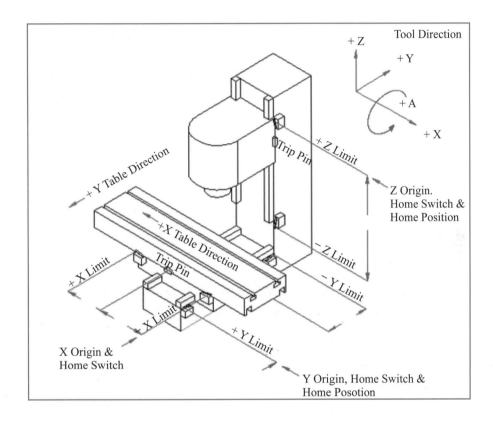

位置（**Positions**）：位置由相對於固定基準的座標系來定義，基本上是 X, Y, Z。每一個工件的位置都由它自己的工作座標所定義。A position is defined by coordinates—typically X, Y and Z—that are relative to a fixed datum. Each position on the work-piece is uniquely defined by its coordinates.

工作平面（**Working-plane**）：工作平面乃所需加工輪廓的平面，一般來講刀具垂直於工作平面。This is the plane in which the contour of a workpiece is being machined and the tool is usually perpendicular to the working plane.

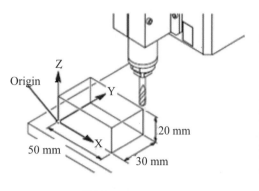

位置 Positions©

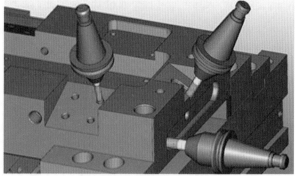

工作平面 Working plane©

建立在工件上的座標系（Coordinate System on Workpiece）

產生 CNC 加工程式以前，先要在工件上定義一個基準。然後就可以通過 CAM 軟體中的軌跡函數及座標系來確立所需加工之工件的輪廓。此參考系稱為工件座標系。Before you create a machining program, you specify a datum to which your coordinate entries for your workpiece refer. You can then specify the workpiece contour in the machining program via tool-path functions and coordinates. This reference system is referred to as the workpiece coordinate system

基準（Datum）：為加工程式中所參考的固定點。對於一個工件來講，基準點可以進行偏移。Reference point to which the coordinate information in a machining program refers. The datum can be shifted to any position on the workpiece.

輪廓（Contours）：工件的形狀是由輪廓所組成，輪廓一般又由平直部分和彎曲部分連續而成。The shape of a workpiece is composed of contours. In the sense of NC machining, a contour consists of straight and curved elements that are joined to each other.

工件座標係（Workpiece coordinate system (WCS)）：對於工件，加工者對其座標進行定義。加工程式中的任意座標值均以此為基準。工件的座標原點即是工件的基準。The machine operator defines the workpiece coordinate system. The coordinate entries in the machining program refer to this datum. The origin of the workpiece coordinate system is at the workpiece datum.

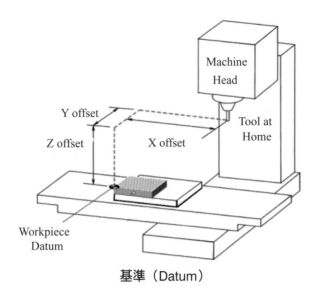

基準（Datum）

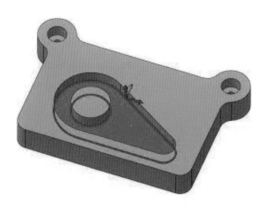

輪廓線（Contours）

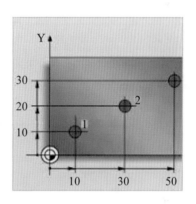

工件座標（Workpiece coordinate system）

二、多軸銑削加工（立式）的座標系基本定義

　　如同大多數加工機，直角座標系是最常用的座標系統。故本書全部以慣用的直角座標系之右手法則進行表述。Cartesian coordinate system is the most used system in machine tools. The invention of Cartesian coordinates in the 17th century by René Descartes (Latinized name: Cartesius) revolutionized mathematics by providing the first systematic link between Euclidean geometry and algebra.

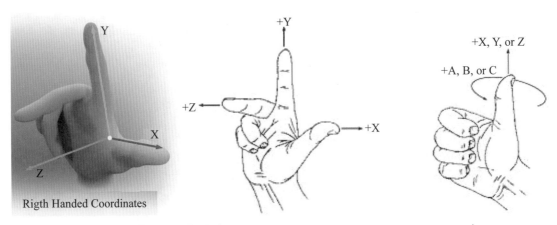

直角座標系之右手法則（Cartesian Coordinate and Right Hand Rule）

　　基於以上規定，**銑削**加工的座標系統就非常的明確，如下圖所示。除了 X, Y, Z 直線運功之外，多軸加工定義三個圍繞三個直線軸的旋轉運動，分別是 A, B, C。Therefore the coordinate system is very clear in milling a work piece。In addition to the three linear movements X, Y and Z, three rotational movements are defined around the linear movement axis, A, B and C respectively, as shown in the pictures.

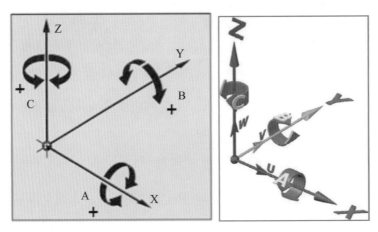

軸向直線及旋轉運動方向規範均服從右手法則（Linea and Rotary Motions）

　　但是在具體的加工機上，運動方向的實際標定常常比較容易混淆，以下是一個立式銑床的範例。However, how to recognize the positive-negtive directions are quite confusion in a real machinetool? Here is the real example：

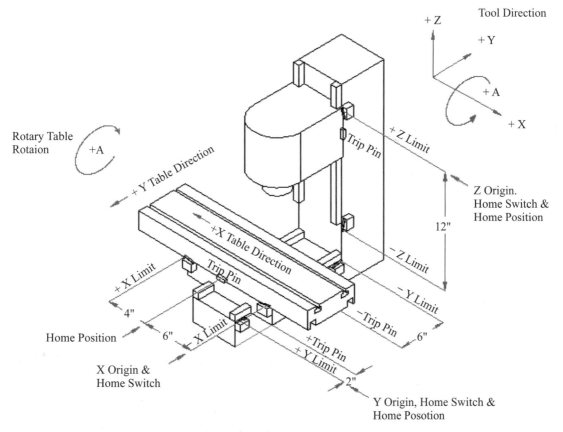

四軸立式銑床部件運動方向解讀（4-Axis vertical milling machine CS）

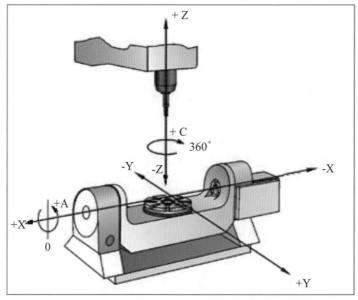

雙旋轉工作臺五軸銑床的軸向規定（Direction of motion in a 5-Axis Table-Table machine）

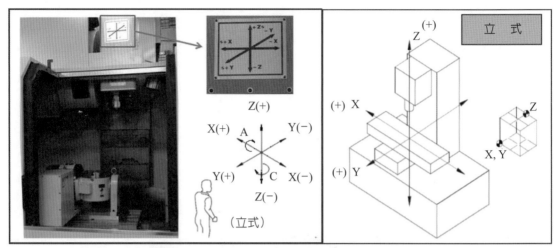

實際 Table-Table 五軸銑床的軸向正負號規定

（Motion directions in a real vertical 5-Axis CNC milling machine）

　　明眼的讀者也許會注意到，以上圖中的有些運動／轉動方向怎麼「反」了呢？重點在於「相對運動」：首先以刀具為主，依以上所示的右手法則確定運動／轉動方向，然後根據機臺上實際運動部件屬於刀具或工件來確定正負。刀具做動的話方向不變，而凡屬於工件運動／轉動的，則取反方向。Above pictures demonstrate the definition of motion directions in a real vertical 5-Axis CNC milling machine, where X, Y, A, C are in "opposite" directions, because of these are the movements of work-piece.

參考文獻（References）

1. http://content.heidenhain.de/presentation/elearning/EN/index.html (HEIDENHAIN)

2. http://en.wikipedia.org/wiki/Cartesian_coordinate_system

3. https://tw.knowledge.yahoo.com/question/question?qid=1612081306728

4. https://www.google.com.tw/search?q=cnc+machine+coordinate+system&biw=1440&bih=785&tbm=isch&tbo=u&source=univ&sa=X&ei=AcNSVZuPKcLQmAWX1oHwAg&ved=0CBsQsAQ#imgrc=l9T3CxTmooLXuM%253A%3Bmv4jqT_7DvuJVM%3Bhttp%253A%252F%252Flinuxcnc.org%252Fdocs%252Fhtml%252Fcommon%252Fimages%252Fmilldiagram.png%3Bhttp%253A%252F%252Flinuxcnc.org%252Fdocs%252Fhtml%252Fcommon%252FUser_Concepts.html%3B640%3B480

5. https://www.google.com.tw/search?biw=1440&bih=785&tbm=isch&sa=1&q=5+axis+cnc+ma

chine+coordinate+system&oq=5+axis+cnc+machine+coordinate+system&gs_l=img.12...1842
888.1845538.0.1847920.7.7.0.0.0.0.33.207.7.7.0.msedr...0...1c.1.64.img..7.0.0.w9RySG3fVG
E#imgdii=Ccr4CE3ZCpa3jM%3A%3BCcr4CE3ZCpa3jM%3A%3B0rQR8i5LBWzpYM%3A
&imgrc=Ccr4CE3ZCpa3jM%253A%3BO5jIjkcw2pZvsM%3Bhttp%253A%252F%252Fwww.
mfg.mtu.edu%252Fcyberman%252Fmachtool%252Fauto%252Fnc%252Fcoordinate.
gif%3Bhttp%253A%252F%252Fwww.mfg.mtu.edu%

6. https://www.youtube.com/watch?v=AKwlzIJG5lo

7. http://content.heidenhain.de/presentation/elearning/EN/index/1271254390233/1271254390247
/1271254390249/1271254390249.html

8. Robert Gian, Chin-Kuo Wang, "Machining with 6-Axis Articulated Robots", JOURNAL OF
LEE-MING INSTITUTE OF TECHNOLOGY, VOL. 23, NO.2, JULY 2012

9. http://en.wikipedia.org/wiki/Cartesian_coordinate_system

10. SINUMERIK 基礎編程手冊，西門子公司，2014

11. 林子寬老師，2013 博士論文，國立台灣大學機械工程學研究所

3

多軸銑削加工刀具軸向設定
（Tool Axis Setup for 5-Axis Milling）

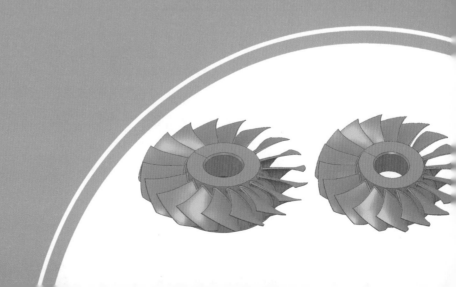

3.1 簡介（Brief Introduction）

對於機台主軸或工作台同時需要進行線性運動和旋轉軸運動的**五軸**加工，**PowerMILL** 提供了多個有效的**刀軸調整**方法和**加工工法**。For 5-Axis milling, where the spindle and work-piece move and rotate separately or simultaneously, several efficient tool-axis adjustment and toolpath strategies are provided by PowerMILL.

　　五軸加工的優點是可通過一次裝夾加工完畢，若使用三**軸**加工需多次裝夾才能加工完零件。可使用**五軸**控制器來重新調整定位刀具，使刀具能沿 **Z 軸**下切到三軸加工方法無法直接加工的口袋深型腔底部或倒勾形面區域。**五軸**加工時，除進行常規的過切檢查外，系統還提供了多個額外選項，確保不同工法路徑在機台、主軸或刀具與加工零件不會發生干涉。進行五軸加工編程時，任何情況下都必須對產生的路徑進行十分仔細的檢查。During 5-Axis Milling, not only regular overcutting-examinations, but also extra options are necessary to avoid collisions between machine/tools and work-piece. Therefore, toolpaths have to be examined very carefully to avoid any collision between tool/shank/holder and workpiece.

3.2 刀軸設定與加工選項
（Tool-Axis Setup and Toolpath Strategies）

PowerMILL 刀軸內定值為**垂直方向**，供三**軸**加工使用。其他選項僅對具有多軸授權的使用者有效。In PowerMILL, the default tool-axis is vertical for 3-axis milling. Other options are only for authorized multi-axis users.

　　刀軸方向選單可透過點擊主工具列中的**刀具軸向** 圖型按鈕，也可直接從五軸**加工工法**選單中選用。To setup tool-axis option, click "Tool-Axis" button in the main tool-bar and can also select directly from 5-aixs tool bar.

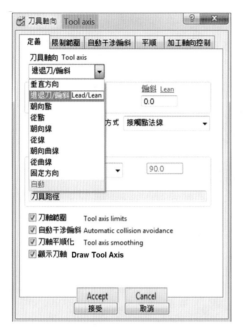

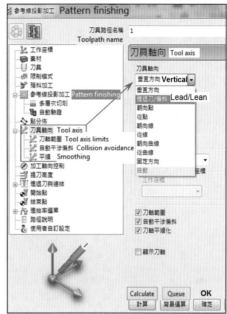

3.3　進退刀／偏斜概述（The concept of Lead/Lean）

　　進退刀角度為刀具沿刀具路徑方向的前後指定角度；**偏斜**角度則是刀具路徑方向垂直左右指定角度。如果這兩個角度的設定均為零，則刀具方向將為刀具路徑的法向。刀具路徑的**法向**為刀具路徑產生過程中，將其投影到曲面資料上時的方向。對**參考線精加**工而言，此方向始終為垂直的；對**投影精加**工而言，其方向隨投影方向的改變而改變。**Lead** allows the tool to be aligned to a specified angle **along** the **toolpath direction** and **lean** a specified angle **across** the **toolpath direction**. If both angles are zero, the tool will be aligned along the direction that a toolpath is projected onto the model during creation. For "**Pattern finishing**" this will always be vertical and for "**Projection Finishing**" it will vary depending on the defined strategy projection options.

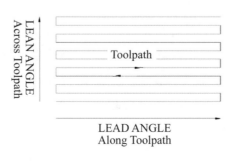

- 全部刪除並重設選單參數（Delete all and reset all parameters）
- 建立素材並依照下圖輸入數值（Setup block and input parameters in the picture）

- 重設提刀高度和開始和終止點選單。Reset "Rapid move height" and "Start/End points".
- 右擊樹狀列中的**模型**選項，從彈出功能表選取**建立平面－自素材**，在 **Z** 高度為 **0** 處產生一平面。Right-click "Model", select "Create Plane" and "From Block", to create a plane at z = 0.

- 建立直徑為 **D5**，長度為 **25** 的球刀，Create a Ball Nosed Tool with 5mm diameter, 25mm long, Name: BN5。
- 建立平行投影加工工法，設定公差為 0.02，**預留量為 0，刀間距為 5，角度為 0，樣式－雙向，長／短連結－相對值**，並將該刀具路徑重新命名為 BN5-Vertical 。Create toolpath, Tolerance 0.02mm, Thickness 0, Step-Over 5mm, Angle 0, Name: **BN5-Vertical.**
- 計算並取消選單。Calculate to get tool path.
- 刀具路徑**動態模擬**。Toolpath dynamic simulation.

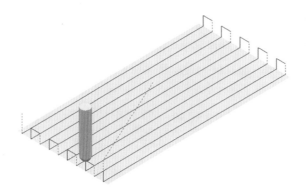

在此我們產生刀具垂直於加工平面的平行刀具路徑，The toolpath is perpendicular tk the plane.

- 右擊樹狀列中的**刀具路徑 BN5-Vertical**，從功能表中選取**設定參數**，打開原始的平行**投影加工**選單。Right-click toolpath "BN5-Vertical", click "Set Parameters",

- 複製 此刀具路徑並將它重新命名為 **BN5-Lead30**。Copy the toolpath and rename as: BN5-Lead30.

- 選取選單中的**刀具軸向**頁面，Select "Tool-Axis" dialog box.

- 定義**刀具軸向**為**進退刀／偏斜**，**進退刀**角度設定為 -30。Define tool-axis direction as "lead/Lean", where "Lead angle" as -30°, and "Lean angle as 0".

- 接受此刀軸向選單（Accept this Tool-Axis Orientation）
- 計算刀具路徑（Calculate the tooplpath）
- 模擬刀具路徑（Simulate the tooplpath）
在此我們產生了刀具於加工平面成一個夾角的平行刀具路徑（The tooplpath is angled with the plane）

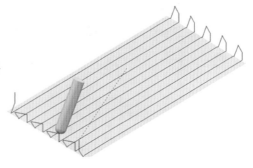

在此我們產生了一平行刀具路徑，**刀具軸向**沿刀具路徑呈 **-30°** 傾斜。此路徑使用**雙向**選項可使**刀具軸向**在每條路徑末端自動改變方向。This way a parallel toolpath is created, there is -30° inclinations. If "Due direction" is selected, the tool-axis will switch direction at each end of the toolpath.

- 利用滑鼠右鍵點選樹狀列中刀具**路徑 BN5-Lead30** ，從彈出功能表選取**設定參數**選項，打開平行投影加工選單。Click toolpath "BN5-Lead30", select "Set parameters", open "Parallel Projection".

- 點擊**編輯現有刀具路徑** ，從打開選單中將**樣式**由**雙向**改變爲**單向**。**計算**並**取消**選單，Click "Edit Existing **Toolpath**", Change "**Both Direction**" into "**Single Direction**". Then calculate and exit.

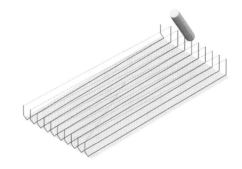

可見，將**樣式**設定爲**單向**後，刀軸方向始終保持不變。The tool-axis direction still keeps the same.

- **右擊樹狀列中的刀具路徑 BN5-Lead30** ，從彈出功能表中選取**設定參數**選項，打開刀**具路徑**選單。Right-click toolpath "BN5-Lead30" and select "Set Parameters".

- **複製** 刀具路徑並將它**重新命名**爲 **BN5-Lean45**。Copy toolpath and rename as "BN5-Lean45".

- 選取刀具軸向選單，Select "**Tool-Axis**"。

- 定義刀軸爲**進退刀 / 偏斜**，**進退刀**角度設定爲 **0**，**偏斜**角度設定爲 **45°**。Set "Lead angle" at 0°, and "Lean Angle" as 45° .

- **接受**此**刀軸方向**選單 Accept the **tool-axis** setting.
- **計算**刀具路徑並**取消**選單 **Calculate** toolpath.
- **動態模擬**刀具路徑 Dynamic simulation the toolpath.

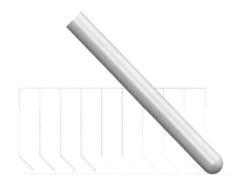

從左 -X 查看 View from left

　　在此我們產生了一刀軸方向**偏斜**於刀具路徑方向 **45°** 的**平行**刀具路徑。This way, the tool-axis has 45° inclination to the toolpath.

　　如果在選單中直接應用了**雙向**工法，那麼刀軸將會在相反路徑方向做左右方向偏斜。If "Both direction" is selected, the tool-axis will switch directions at each end of the toolpath.

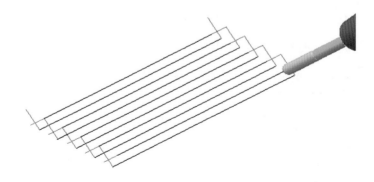

可通過編輯一個**單向**刀具路徑來產生一具有恒定**偏斜**方向的**雙向**工法，A "Two-way"

toolpath with constant lean direction can be created by editing the "Single-way" toolpath.

通過在**樹狀列**中**右擊**相應**刀具路徑**，從彈出功能表選取**編輯 - 重新排列**選項，從彈出選單中點擊**單雙向切換**  來修改單向工法。原始刀軸方向將不受影響，保持不變。Right-click the toolpath, select "Edit-Rearrange", select "Single-Double transfer" to edit the toolpath. Original tool-axis will not be affected.

- 右擊**樹狀列**中的**刀具路徑**，從彈出功能表選取**編輯 - 重新排列**選項，打開**刀具路徑**列表選單。Right-click the toolpath, select "Edit-Rearrange", open toolpath list.

- 點擊**單雙向切換** ，我們可看到正如 **PowerMILL** 警告方框中所說的，**刀軸**方向維持同樣方向做加工。Click "Single-double transfer" , as you will see in the warning window: Machining with same tool-axis direction.

提示 Note：以路徑前進方向來檢視，偏斜正角度為左偏斜，負角度為右偏斜；進退刀正角度為向前，負角度為向後。Viewing from toolpath forward direction, positive lean angle is toward the left, negative angle is to the right; while positive lead angle is toward the front, negative angle is to the back;

3.4　範例 -1 進退刀 / 偏斜，朝向點 / 從點（Example-1 Lead/Lean, Touard/From Point）

- **全部刪除**並**重設選單參數**，Delete all and reset parameters.
- 經由光碟 Chapter-03 **輸入模型**，Import model "**joint5axis.dgk**".

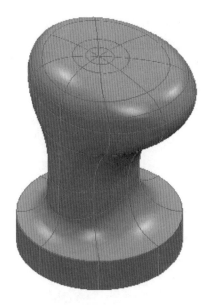

- 依物件最小／最大值產生素材並將素材在 **X** 軸和 **Y** 軸方向**延伸 15mm**。Create block material, calculate and then extend 15mm in both X and Y directions.
- 定義一球刀直徑 Define **Ball Nosed Tool** of **25mm** diameter, Name: **BN25**。
- 點擊提刀高度選單中的**計算**，Setup "**Rapid Move**".
- 在**開始點**選單中設定**素材中心安全高**，**結束點**選單中設定**最後一點安全高**。
- 按以下參數修改**進退刀與連接** ，Setup "Lead and Link" parameters：

Z 軸高度，Z axis height：相對 **Relative** 提刀高度 45，緩降高度 **Slowly down height** 10

　　　　　　連結 Link：增量值 Incremental

- 選取**工法**選單 Toolpath strategy ，在選單中選取**精加工** Finish。
- 開啟**線投影加工**和**刀具軸向**選單，如下圖所示，依序將數值輸入。

  Select "**Projection Line Finishing**" and **Tool-Axis**, input the following values.
- **設定方位角**，Setup **Azimuth** at 0°; 設定仰角，Setup **Elevation** at 0°

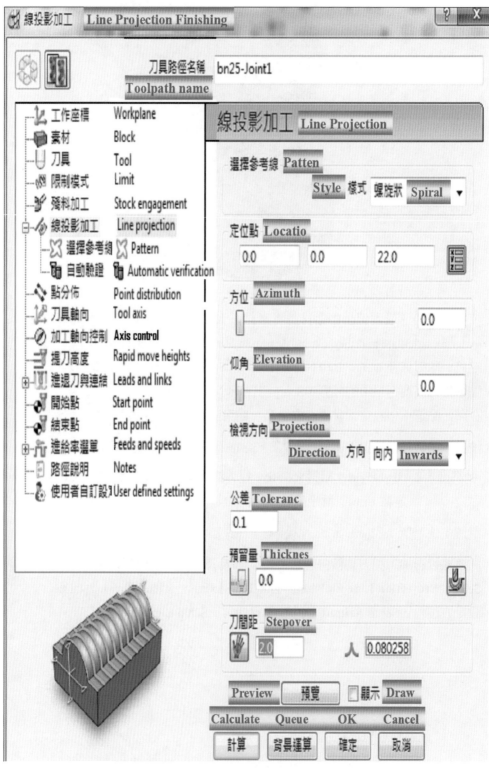

中 - 英對照加工對話框（Chinese-English Toolpath Setup Table）

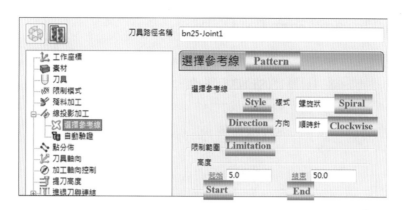

- **計算**此刀具路徑 Calculate toolpath，然後**取消**並關閉選單 Close the dialog box。

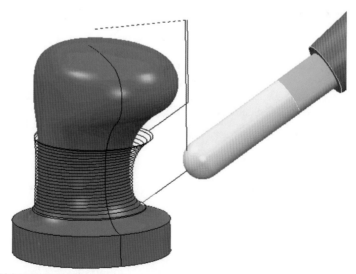

註 Note：下面將繼續加工此零件的上半部分，The following is to machine the top part.

五軸銑削數控加工之基礎及實作

朝向點／從點（Toward / From Point）

　　此選項允許產生五軸加工刀具路徑的過程中，基於使用者定義的點來定位**刀具軸向**。刀軸的實際方向是相對於刀具路徑的**預覽**投影線。**朝向點**選項適合於加工外部形狀（如凸出物），而從**點**選項適合於加工內部形狀（如口袋）。此模型上半部分，適合以**朝向點**選項進行加工。This option is to align the tool-axis orientation based on user defined point.

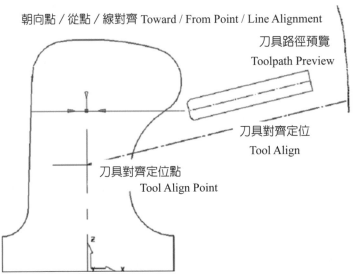

朝向點／從點／線對齊 Toward / From Point / Line Alignment
刀具路徑預覽 Toolpath Preview
刀具對齊定位 Tool Align
刀具對齊定位點 Tool Align Point

註 Note：上圖的刀具軸向同樣應用於 - 朝向線／從線，Above tool-Axis orientation is also applied to Toward Line/From Line.

- 按照下面給定參數設定**進退刀與連接選單** Setup Lead/Link parameters：

- Z 軸高度 Z Height：相對提刀高度 Skim Dist. 45，緩降高度 Plunge Dist. 10
- 進刀 Lead in：水平圓弧 Horizontal Arc，半徑 Radius 6，角度 Angle 90
- 退刀 Lead out：垂直圓弧 Vertical Arc，半徑 Radius 6，角度 Angle 90
- 路徑延伸 Extension：延伸移動 Extended move
- 連結 Link：增量值 Incremental

- 點擊工法選單 Click toolpath，在選單中選取**精加工** Select Finish machining。
- 開啟**點**投影加工和刀具軸向選單，如下圖所示，依序將數值輸入，Select "Project Point Finish" and "Tool-Axis". Input parameters as shown in the following pictures.

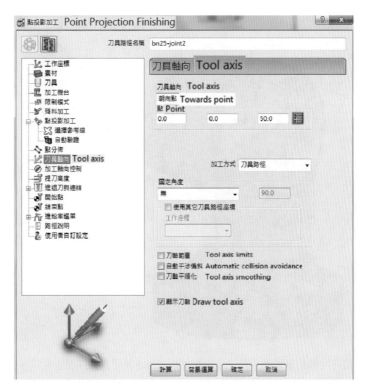

- 設定仰角 Setup **Elevation**：開始 Start at 90° - 結束 End at 0°

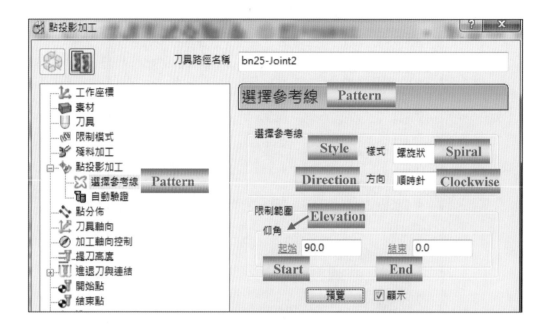

- **計算**此刀具路徑，然後**取消**並關閉選單，Calculate toolpath and close dialog box。

註 Note：刀具定位點位在投影加工範圍焦點之下，約 10mm。這樣可確保加工過程中，主軸相對於工作平台有一定仰角，進而避免干涉碰撞。The tool point is to set under projection machining about 10mm. This setup is to assure there is an elevation angle and avoid collision.

3.5　範例 -2 朝向線 / 從線 （Example-2 Toward / From Line）

該選項允許五軸刀具路徑，透過使用者自訂的直線來定義刀具**軸向**。直線透過 **XYZ** 座標位置，以**向量**方式定義。**朝向線**選項適合於加工外部形狀（如凸出物），而**從線**選項適合於加工內部形狀（如口袋）。This option is to align the tool-axis orientation based on user defined lines. The straight lines are defined by vector through XYZ coordinate.

- 全部刪除，**Delete every thing**。
- 經由光碟 Chapter-03 **輸入模型**，Import model **from-line-model.dgk**。

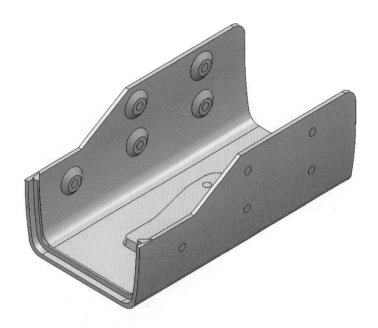

- 產生直徑爲 **12mm**　，長度爲 **55**，刀桿直徑 **12**，長度 **40**，第一夾頭底部直徑 **25**，頂部直徑 **40**，長度 **40**；第二夾頭頂部 / 底部直徑 **40**，長度 **60**，伸出長度 **90** 的球刀，**Create a Ball Nosed Tool:**

  (a) Tool: 12mm diameter, 55mm length;

  (b) Shank: 12mm diameter and 40mm long;

  (c) First Holder: 25mm bottom and 40mm top, 40mm long;

  (d) Second Holder: 40mm bottom and 40mm long.

- 依**物件最小 / 最大值**定義**素材**，Make a block material based on model dimension.
- 點選**提刀高度**選單中的**計算**，Click "Calculate" on "Rapid Move Height".
- 同時將**開始和終止點**設定爲素材中心安全高，Set "Start/End" at centre of block.

- 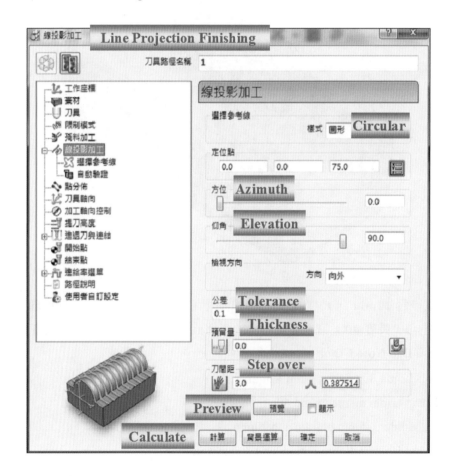 設定相對提刀高度和緩降高度為 **5**，Set Rapid and Plunge clearances all as 5;
- 全部進退刀和路徑延伸為無，Set Lead in/out and toolpath extension as "no";

  連結－長短連結分界值 : Link – Short/Long threshold: 6;

  > 短連結：**圓弧**，Short link: Arc;

  > 長連結：**增量值**，Long link: Incremental;

  > 預設值：**增量值**，Pre-set value: Incremental;

- 點擊工法選單按鈕 ，在選單中選取**精加工**選項，Open Finishing machining window.

- 開啟**線投影加工**和刀具**軸向**選單，依照下圖所示，依序將數值輸入，Select "Point Projection Finishing" and input the following values. 設定方位角 Set Azimuth angle at 0°; 設定仰角 Set Elevation angle at 90°

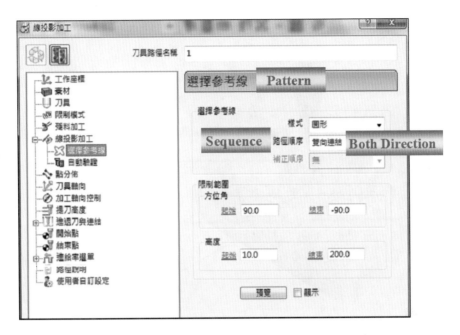

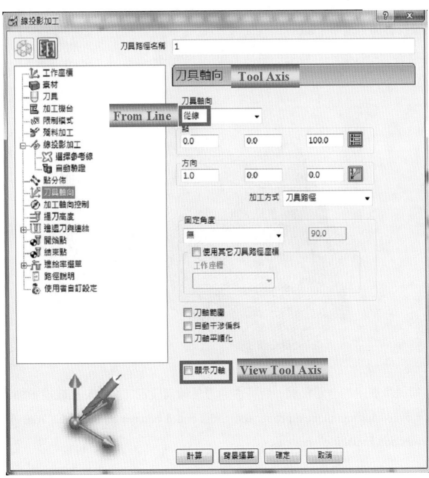

• 勾選**顯示刀軸**方框，顯示相對於模型上的刀軸方向，Check "View Tool-Axis"

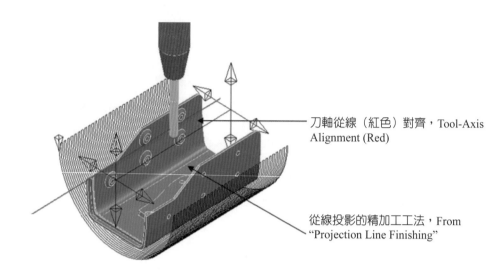

刀軸從線（紅色）對齊，Tool-Axis
Alignment (Red)

從線投影的精加工工法，From
"Projection Line Finishing"

• 點擊**預覽**按鈕，查看工法，最後點選**計算**按鈕，Preview, confirm and calculate.

• 關閉**線投影加工**選單，Close the dialog box.

下圖是最後得到的結果，The result is as the following。

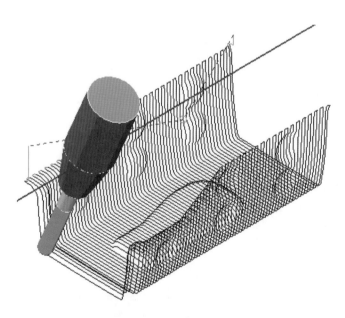

包含倒勾區域在內的全部形狀，皆可透過**刀具軸向**設定（從線）和**線投影加工**產生適當的路徑，For over-hanging shapes, proper toolpaths could be created through "Tool-Axis" setting and "Line Projection Finishing" strategies.

3.6　範例 -3 朝向曲線 / 從曲線 （Example-3 Toward/From Curve）

此選項允許五軸刀具路徑，依照使用者定義曲線（**參考線**）作為**刀具軸向**的依據。

註：下面章節將針對此範例的**沿面投影加工**再做更詳細的介紹。This option is to align the tool-axis orientation based on user defined curves (Patterns). Following is the example of "Surface Projection Finishing".

- **全部刪除**，Delete all.
- 經由光碟 Chapter-03 **輸入模型**，Import model: **impeller.dgk**。
- 產生一空白的**參考線**並將它重新**命名**為 **Align2Curve**，Create an empty Pattern and rename it as **Align2Curve**.。
- 從工作視窗中選擇此 curve 軸向的參考線，In the graphics area, select the wireframe
- 從參考線名稱（Align2Curve）右鍵功能列中，點選插入 > 模型檔案 > 經由經由光碟 Chapter-03 選擇 "**Align2Curve.dgk**". Right click pattern curve (Align2Curve), insert pattern file from CD Chapter-03 **Align2Curve.dgk**, the alignment curve as a Pattern segment.

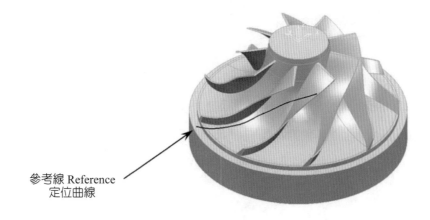

參考線 Reference
定位曲線

- 產生**直徑**為 **3，長度**為 **15** 的**球刀 BN3**，刀桿和夾頭資料如下，Creat a ball nosed tool, with 3mm diameter and 35mm length, Name:BN3.
  刀桿，Shank：頂部 / 底部直徑 3，長 25，top/bottom 3mm diameter and 25mm long
  夾頭 1，Holder1：頂部直徑 15，底部直徑 10，長度 50，Holder1, top 15, bottom 10, length 10mm
  夾頭 2，Holder2：頂部 / 底部直徑 15，長度 35，伸出長度 25，top/bottom 15, length 25mm
- 依 … 定義選擇圓柱產生一**圓柱體素材**，Create cylindrical block material.
- 設定進刀 / 退刀為**垂直圓弧**：**距離 0，角度 90，半徑 3**, Set Lead/lean as "Vertical arc": Distance 0, angle 90, radius 3mm.

連結－長短連結分界值：**6**; Link – Long & Short link conjunction: 6;

短連結：**圓弧**，Short link: Arc;

長連結：**增量值**，Long link: Incremental;

預設值：**增量值**，Pre-set value: Incremental;

- 點擊提刀高度選單中的**計算**，Click "Calculate" for "Rapid Move Height" for the tool.

- 將**開始點**和**結束點**均設定為**素材中心安全高**，Set "Start" and "End" points at the safe height of block centre.

- 選取位在**參考線**旁上邊葉片的背部**曲面**，Select the back surface of blade surface next to the reference curve,

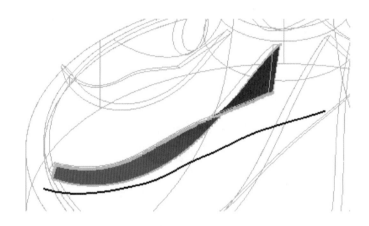

- 點擊**工法選單按鈕** ，於選單中選取**精加工**頁面，Go to Finishing window.

- 開啟**曲面法向投影加工**和**刀具軸向**選單，依照下圖所示，依序將數值填入，Select "Surface Finishing" and "Tool-Axis", fill the values as shown in the following pictures.

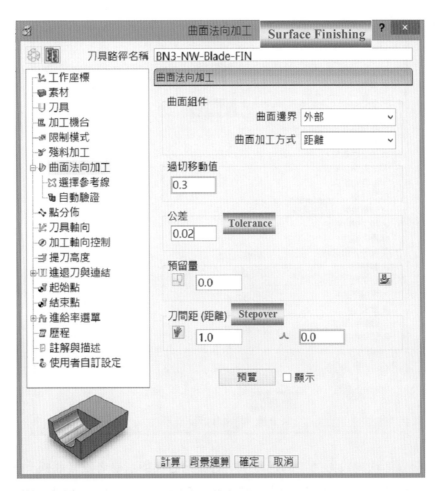

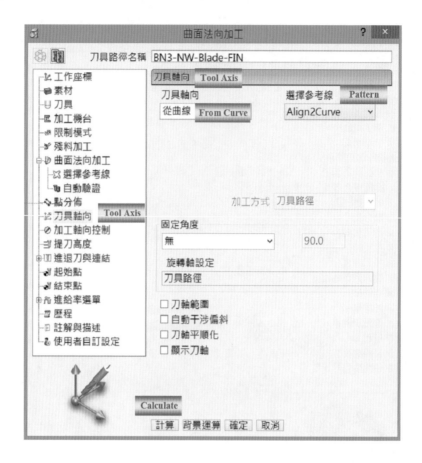

透過模擬曲面法向投影加工的過程中，刀軸始終和所選**參考線**（曲線）對齊，From "Surface Finishing", tool-axis always aligns with the reference curve, shown as the following:

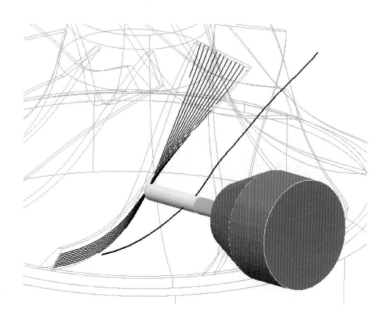

- 使用滑鼠右鍵點選**曲面法向投影加工工法**，並選取**設定參數**，重新開啟選單，Right-click "Surface Finishing", select "Set parameters".
- 使用滑鼠左鍵點擊選單中的**複製刀具路徑按鈕** ，Click "Copy Toolpath".
- 選取位在**參考線**旁下邊葉片的上部**曲面**，Select higher part of blade surface next to the reference curve.

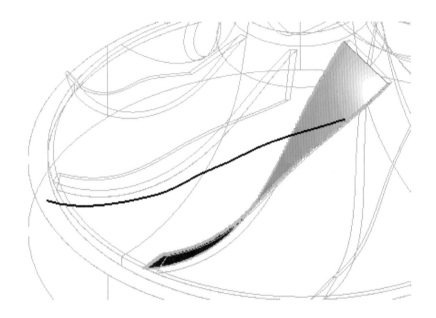

- 開啟**曲面法向投影加工**和**刀具軸向**選單，依照下圖所示，依序將數值填入，Select "Surface Finishing" and "Tool-Axis", fill the values as shown in the following pictures.
- 刀具路徑名稱，Change toolpath name - BN3-**UP**-Blade-FIN

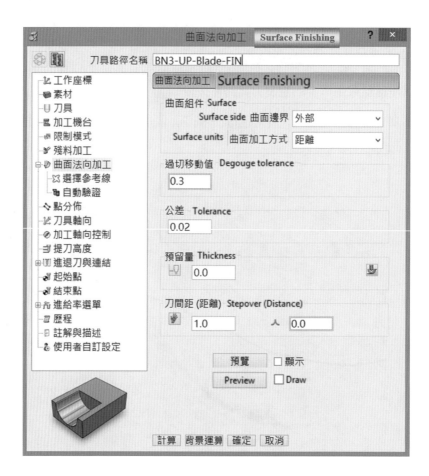

- 參考線方向，**Pattern direction–V**,
- 起始角落 **Start corner** - 最小 **Minimum U,** 最大 **Maximum V**
- 點擊預覽按鈕，查看工法預覽，最後點擊計算，產生刀具路徑，Preview and Calculate to get the tool path.
- 關閉此沿面投影加工選單，Close the dialog box.

使用者模擬所選加工工法，可以發現刀軸始終和參考線（曲線）對齊。With "Surface Finishing toolpath strategy", tool-axis always aligns with the reference curve, shown as the following picture:

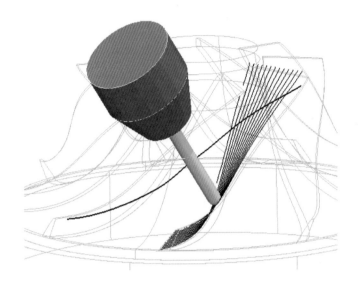

3-7　範例 -4 固定方向（Example-4 Fixed Direction）

此選項允許使用者，將刀具軸向設定為**指定的向量角度**方向。下述範例將使用該選項，針對模型上傾斜凹陷的部分進行精加工。This option is for the tool-axis to the fixed vector-directions.

向量和角度轉換表（Table of transformation between angle and vector）

固定方向的刀具軸向設定 — 向量 / 角度對應關係，下表列出了角度方向和 XY 平面（Z = 0）向量之間的關係，The following table are the transformation between angles and vectors to the XY plane (Z = 0).

角度，Angle（度）	向量，Vector（I J K）		
0	1	0.0000	0
5	1	0.0875	0
10	1	0.1760	0
15	1	0.2680	0
20	1	0.3640	0
25	1	0.4660	0
30	1	0.5770	0
35	1	0.7000	0
40	1	0.8390	0
45	1	1.0000	0
50	1	1.1920	0
55	1	1.4280	0
60	1	1.7320	0
65	1	2.1450	0
70	1	2.7470	0
75	1	3.7320	0
80	1	5.6710	0
85	1	11.4300	0
90	0	1.0000	0

$$\text{Tan (Angle)} = \frac{\text{(opposite)}}{1}$$

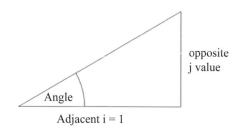

　　透過輸入適當數值，定義刀具軸向（朝向主軸）的 **IJK** 向量。將**刀軸**設定對於當前作動工作座標的一**固定方向**。儘管定義向量值，需要具備基礎的三角函數知識，但這種方法可以十分靈活地定義零件加工角度。Input proper IJK values of the vector, you can easily set a fixed tool-axis orientation to the active work coordinator, even without the basic knowledge of trigonometric function.

- 全部刪除，重設選單參數，Delete all and reset parameters.
- 經由光碟 **Chapter-03** 開啟唯讀專案，Open the read-only project：**ToolAxisFixed1-Start**

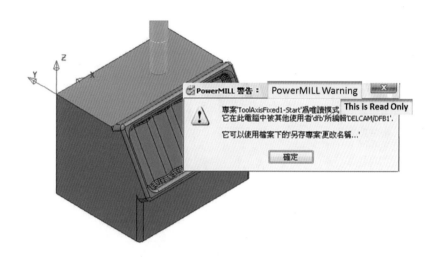

- 另儲存專案爲，Save the project as：
  XXX:/PowerMILL_Projects/ToolAxisFixed-EX1
- 作動工作座標，Activate Work Coordinate - **MC-Datum**.
- 作動刀具，Activate Tool - **BN6** .
- 選取傾斜凹面的波狀**曲面**，Select the **wavy surface** as shown.

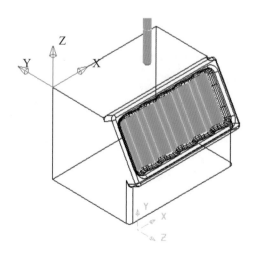

- 點擊工法選單圖型按鈕 ，在工法選單中選取**精加工**頁面，Open "Finishing" toolpath window.
- 選取**曲面法向加工**選項，請按下圖所示依序填入相關數值，Select "Surface Finishing", and input the values in the following pictures.

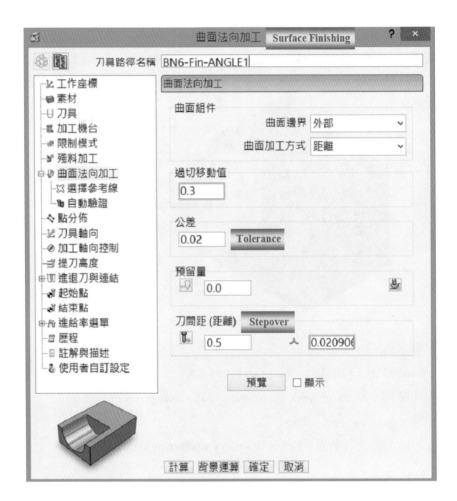

接下來選擇參考線，**Select reference pattern.**

• 參考線方向，**Pattern direction - V**

• 加工順序，**Ordering - 雙向，Two-way**

• 打開**進退刀與連結**選單，並切換至**連結**頁面，Select "Link" in "Lead and Links".

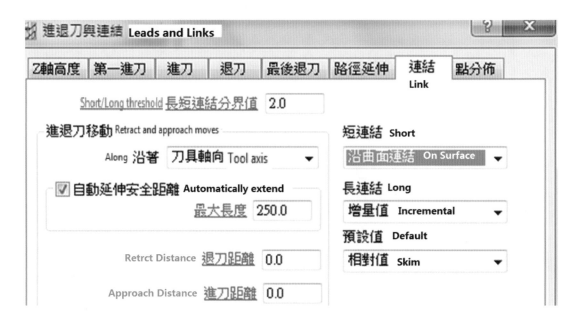

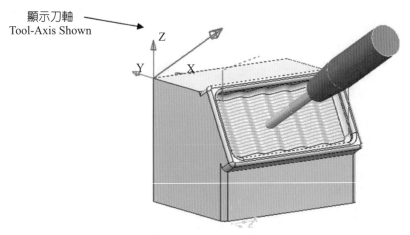

顯示刀軸
Tool-Axis Shown

- 選擇所需加工的曲面,點擊**預覽**按鈕,查看工法,然後點擊**計算**,產生刀具路徑,
  Select the surface, Preview, Confirm and Calculate.
- **關閉此曲面法向加工選單**,Close the dialog box。

註 Note:為方便查看,該條刀具路徑,使用的 2mm 刀間距,For clarity, the "Step Over" is set as **2mm**。

- 從下拉式選單選取**檔案 - 儲存專案**,更新儲存的專案,Save project as:
  XXX:/PowerMILL_Projects/ToolAxisFixed-EX1

四軸加工範例：造型香水瓶
（4-Axis Macnining: A Rotary-Bottle）

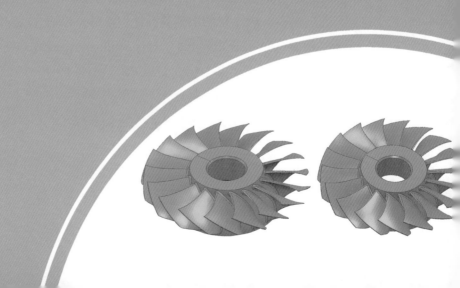

4.1 基本設定（Basic Setup）

一、經由光碟 Chapter-04 輸入模型：（香水瓶模型）Import the model from CD Chapter-04: Rotary-Bottle

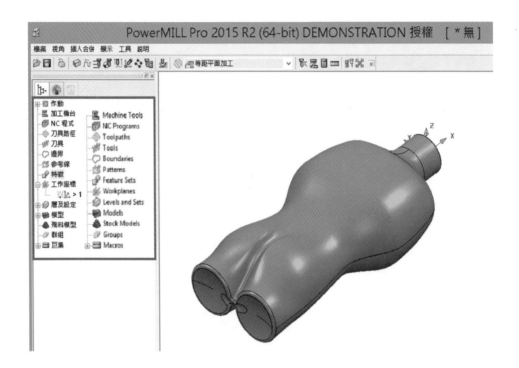

二、工作座標設定 Setup the Work Coordinates

瓶子的頂端在世界座標，X 軸橫向朝右。接下去需要，The top of the bottle is at the origin of world coordinate, with x leads to the right at center. Then we need create:

(1) 建立工作座標 WC1 目的為定義圓柱型素材，Work coordinate WC1 for cylindrical block: Rotate 90° about Y axis from the world coordinate, the Z vector faces the back of the bottle for WC1.

* 注 Note：在這里工作座標和座標平面往往稱為：Work-Coordinate (WC) or Work-Plane are same in this application.

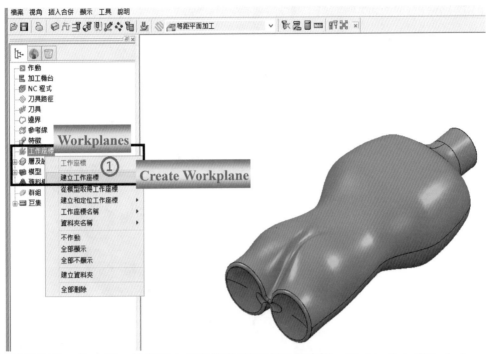

從物件管理列的工作座標，按滑鼠右鍵功能點選建立工作座標，Right click "Workplane" and select "Create Workplane"

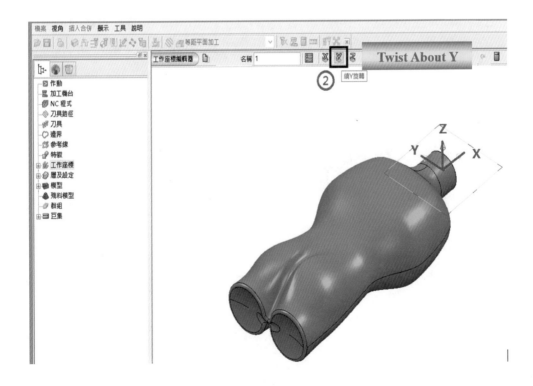

五軸銑削數控加工之基礎及實作

如上圖顯示出現工作座標編輯工具列，點選繞 Y 旋轉功能，Twist 90° about Y. 從旋轉選項視窗中，輸入旋轉角度 90° 接受，再點選 ✓ 完成座標係設定。Input 90° for the angle and accept ✓ .

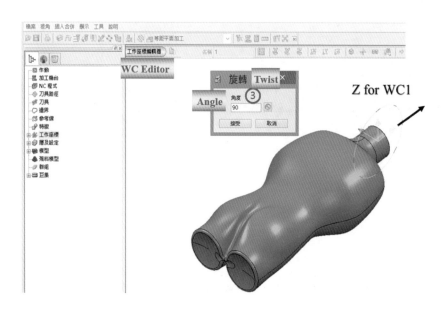

(2) 建立工作座標 WC2 為背面加工，不作動 WC1 工作座標，從物件管理列的工作座標，按滑鼠右鍵功能點選建立工作座標，建立 WC2 工具座標係（可從工具列中重新命名）。

✓ 由工作座標編輯工具列，點選繞 X 旋轉功能，從旋轉選項視窗中，輸入旋轉角度 180° 接受，再點選 ✓ 完成座標係設定。Create work coordinate WC2 for back machining: Rotate 180° about X axis from the world coordinate, the Z vector faces down for WC2.

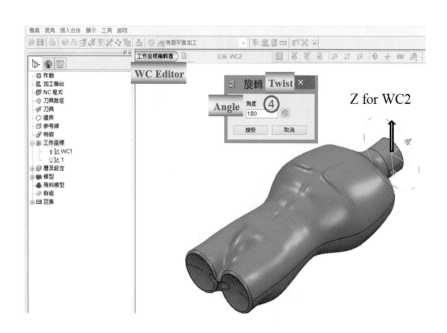

三、素材定義（Define material block）

從主要工具列，點選建立素材 <img>，使用圓柱素材選項，選擇工作座標 WC1，計算後長度座標最大值為 0mm、最小值 -115mm，直徑設定為 ⌀68mm, Select Cylindrical block, with work coordinate WC1, calculate for the dimension. Conduct minor adjustment if necessary.

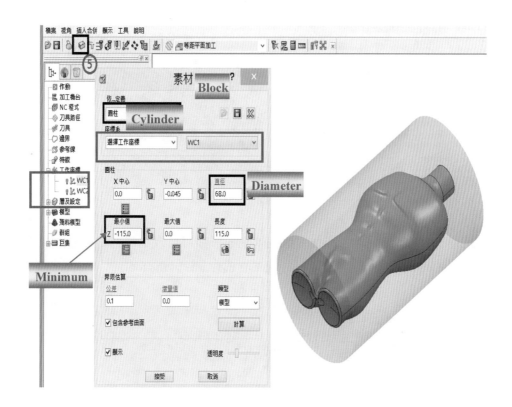

四、刀具設定（Setup for the tools）

為簡便起見，本案例只用一把圓鼻刀，直徑 D8mm，圓角 R2mm。For simplicity as the first example, one cutting tool is used with diameter of 8mm and round corner of 2mm.

從物件管理列的刀具，按滑鼠右鍵功能點選產生刀具，選擇刀具類型圓鼻刀，

Right-click "Tool" and create Tip Radiused Tool.

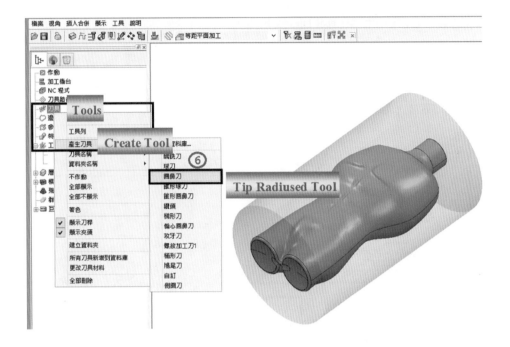

從刀具選項中，設定直徑 D8mm，圓角 R2mm，你可自行定義名稱與刀具號碼。Input tool diameter and corner radius. You can easily define the name of the tool.

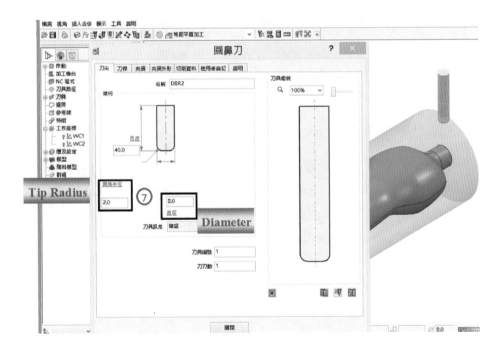

4.2　粗加工（Rough Machining）

一、正面粗加工（Machining for the Front）

- 刀具路徑工法選單 ⬙ ：選擇模型粗加工，Toolpath Stratage: Select Model Area Clearance;

- 選擇世界座標，select World Coordinate;

- 設定提刀高度 ⬙ ，計算，接受，Set and calculate rapid move heights;

- 設定刀間距 = 0.7 刀具直徑 = 5.6mm，Set step over at 0.7×8 = 5.6mm;

- 設定每層下刀 2mm, Set step down at 2mm;

- 設定 Z 軸限制最小值 −2mm, Set machining limitation minimum Z = −2mm.

- 設定進退刀與連結 ⬙ ，進刀 = 斜向下刀 Set lead in at ramp.

- 按計算，得到正面加工的刀具路徑，Calculate and complete the tool path.

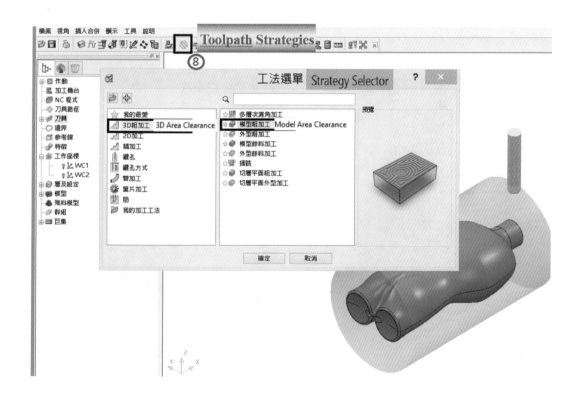

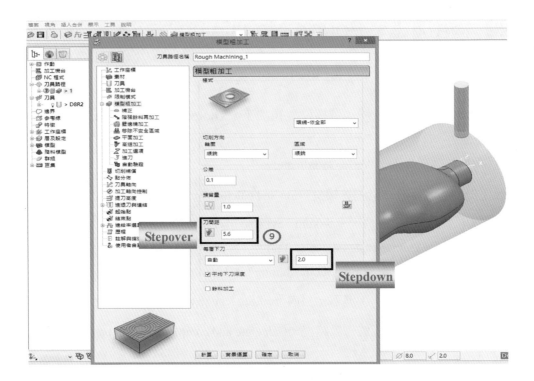

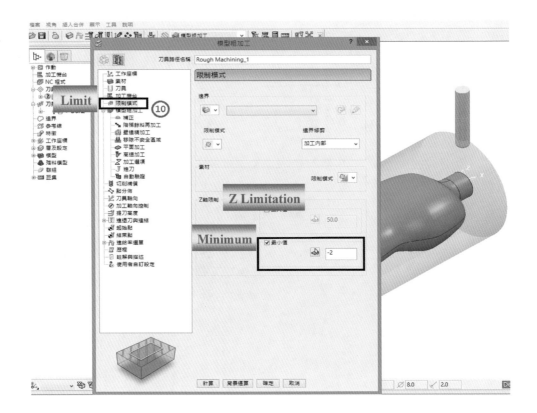

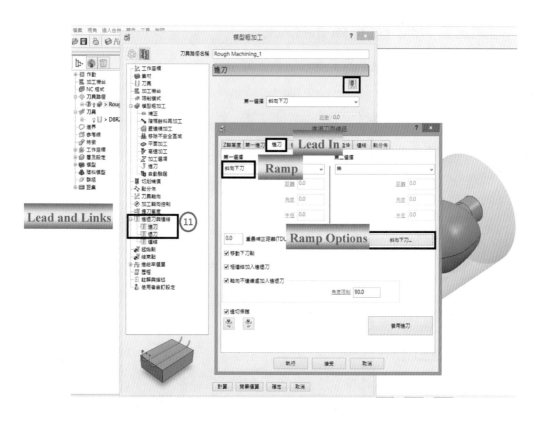

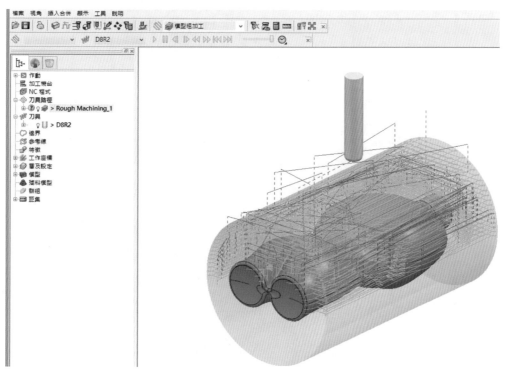

五軸銑削數控加工之基礎及實作

二、背面粗加工（Machining for the back）

- 作動工作座標 WC2, Activate the World Coordinate 2 (WC2)
- 顯示工件實際翻轉後的視角，View the work piece upside down;
- 須注意重新依 WC2，設定提刀安全高度 ，計算，接受，Set rapid move heights;
- 重複正面加工的步驟，得到反面加工的刀具路徑如下，Repeat the steps in front machining to complete.

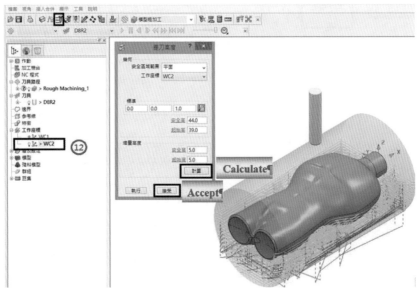

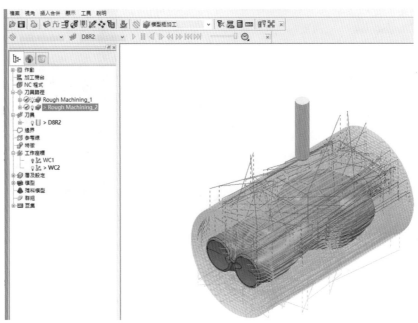

4.3 精加工（Finish Machining）

- 恢復世界座標，Recover the world coordinate by deactivate WC2;

- 選擇 精加工中的「旋轉投影加工」工法；Select "Rotary-Finishing"

- Select rotational-projection-method in finish machining.

- 設定公差值為 0.01，預留量為 0; Set tolerance as 0.01, thickness as 0

- 設定刀間距 0.5mm；設定 X 軸加工範圍 −122mm 至 0mm；Set step over as 0.5mm and machine range from X = −122 to X = 0;

- 消除 Z 軸最小值 −2mm 的限制，Be sure to remove the Z minimum limitation;

- 完成精加工產生刀具路徑，Calculate and complete the finish machining.

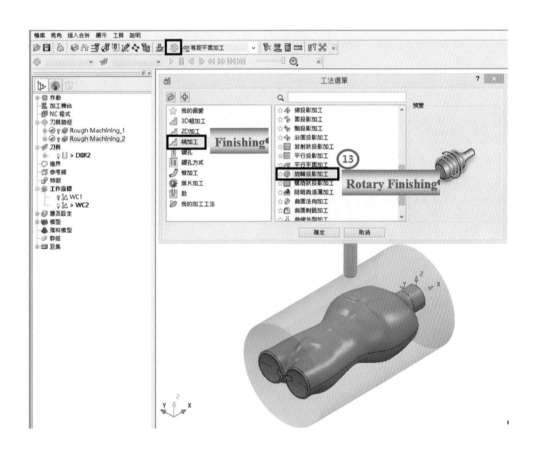

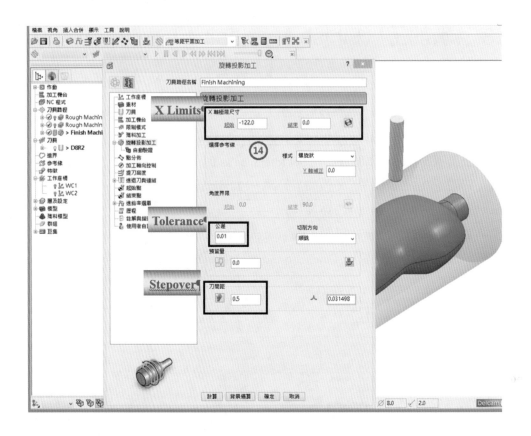

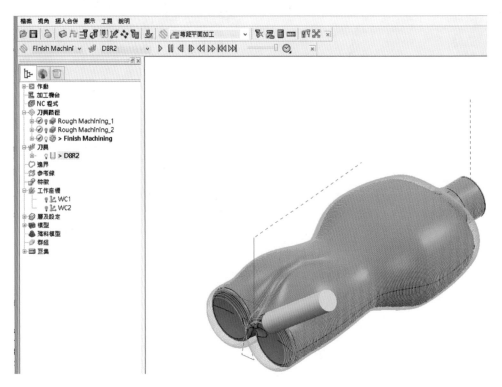

4.4　實體加工模擬（Machining Simulation）

　　開啓實體模擬等兩項工具列，選擇刀具路徑 1 及刀具 1，按 Play 即可觀賞正面粗加工的模擬情況。背面粗加工及精加工模擬步驟相同。記得路徑模擬之前，須按刀具起始位置。Turn on the machining-simulation tools. Select tool path 1, tool 1 and start to play the simulation for front part machining. Continue to view the simulations for the rest of the machine process. Remember before each path simulation, you need to select "Simulate from start".

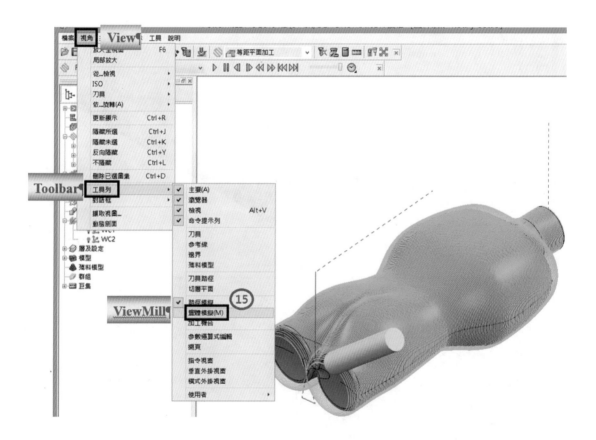

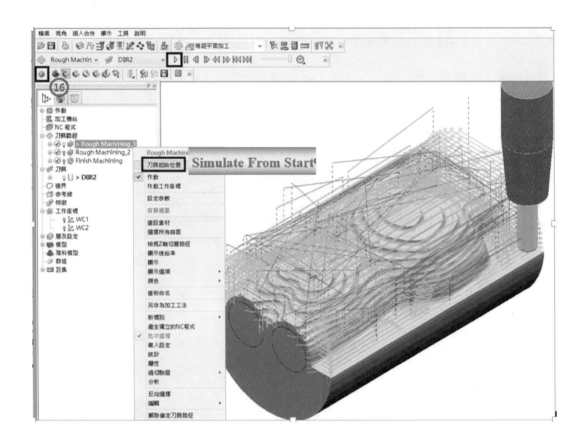

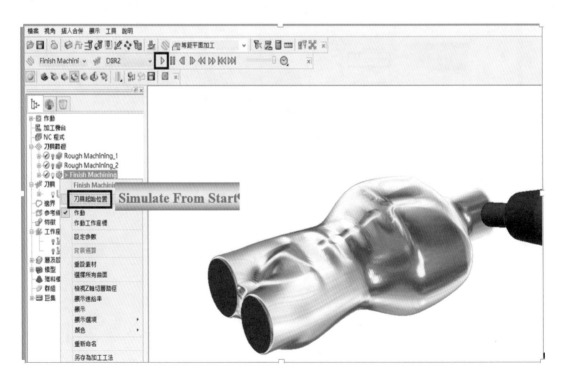

4.5 產生並輸出 NC 碼（Export NC Code）

CAM 軟體產生的刀具路徑程式必須要轉換成 NC 碼，才可以送到加工機上進行加工，
In order to be sent to machine-center, the tool path created with CAM software must be transformed into NC codes:

- 將滑鼠移至「NC 程式」並按下右鍵，在產生的右鍵對話功能中選擇「設定選單」；定
  義儲存使用「專案 - 開」，輸出檔案 NC 副檔名與儲存位置，選擇控制器參數檔。（此
  先行設定可套用到每一條刀具路徑，不必單一逐一設定。）

  Right click "NC Program", Select "Preference" in the dialog-box; Define "Use project" as
  "on"; Select the extension and the location of the NC file; Select the controller file. In this
  way, you have setup for all the tool paths.

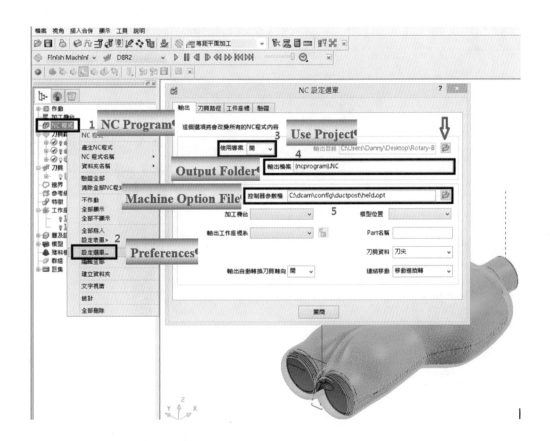

- 選擇所要輸出 NC 碼檔案的位置，select where to export the NC code files
- 選擇控制器參數檔，Select controler file.

以上是根據不同加工機控制器撰寫的輸出程式,其中 "PW-Roland-40A.opt" 是 PowerMill 為了 Roland-40A 加工機所撰寫的轉檔程式,而 "PW-Siemense-40Dsl" 是為了東台精密機械 生產的五軸加工機上西門子控制器 SIMURIK 40D sl 所寫。Above picture lists four opt files specifically writen for particullar controllers from PowerMill. The first one is for Roland-40A 4-Axis milling machine and the second is for Siemense SIMURIK 40D sl controler.

- 將滑鼠移至「刀具路徑」並按下右鍵,在產生的對話方塊裡選擇「產生獨立的 NC 程式」,Right click at "Tool path", and click "Create individual tool path".
- 再從「NC 程式」右鍵功能對話框按下「全部寫入」,即可輸出所需的 NC 碼,Right click at "NC program" and click "Write All", you can export all the NC programs.

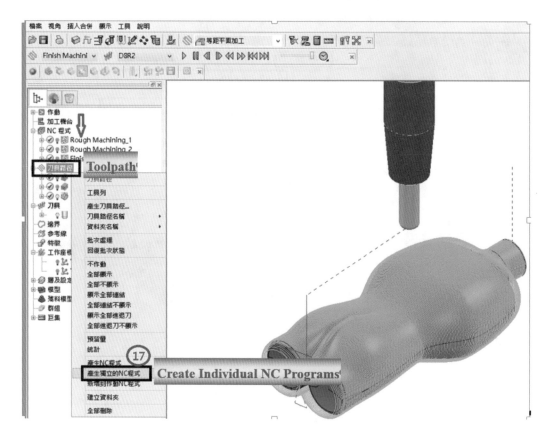

產生個別的 NC 程式,Create individual NC program

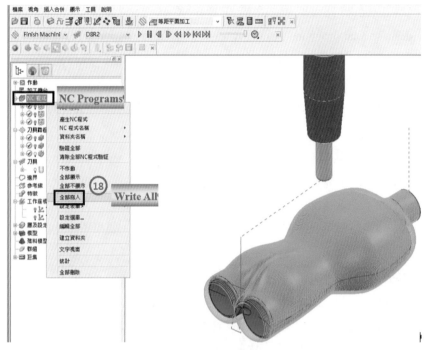

全部寫入，Write All

如要個別輸出單條路徑，可從 NC 程式名稱右鍵功能對話框按下「設定參數」，開啓 NC 程式設定選單，寫入，If you like to export single tool pass, just right click on the NC program and then click "set parameters". On the dialog-box complete setting and "Write".

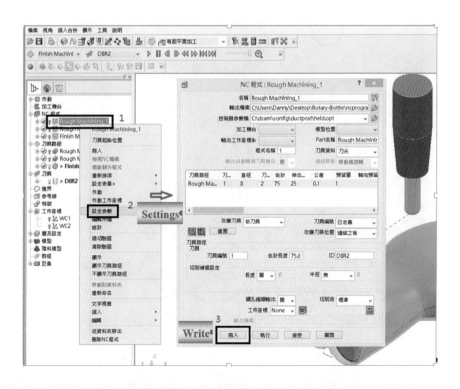

Delcam 公司還提供特別 NC 碼輸出檔 "PostToNC.exe"，以便整體輸出。具體用法如下，
Delcam provides a special program "PostToNC.exe" to make the export more convenient:

1. 啓動 PM 開啓專案 NC 程式要有程式準備輸出，輸出的資料夾需先設定，Under
PowerMill and complete the selections for NC code output shown above;

2. 點擊兩次 "PostToNC (Dongle Locker).exe" 出現對話框，Double click "PostToNC
(Dongle Locker).exe". Then just make the output following the steps:

3. 步驟 Step 1：點擊 Click "Import"

　　步驟 Step 2：選擇要輸出的 Select NC file

　　步驟 Step 3：點擊 Add 加入到右邊對話框準備輸出，Click "Add"

　　步驟 Step 4：點擊 Back 還原到左邊對話框放棄輸出（如果需要的話），If necessary,
　　　　　　　　click "Back" to give up the export action.

　　步驟 Step 5：選擇 POST（於下方顯示所選的 POST 名稱），Select "POST"

　　步驟 Step 6：可輸入欲輸出 NC 的附檔名，Select NC file name

　　步驟 Step 7：點擊 Click「輸出 NC」，Click "Export NC"

　　步驟 Step 8：若輸出的 NC 有 Warning 或 Error 的話，會跳出訊息顯示視窗，If there
　　　　　　　　are Warning or Error, a message box will appear.

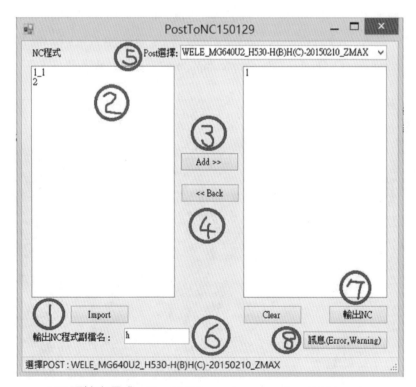

NC 碼輸出程式，PostToNC.exe for exporting G&M code

五軸加工範例：3+2 定面加工（5-Axis Machining: 3+2 Fixed-Face Milling）

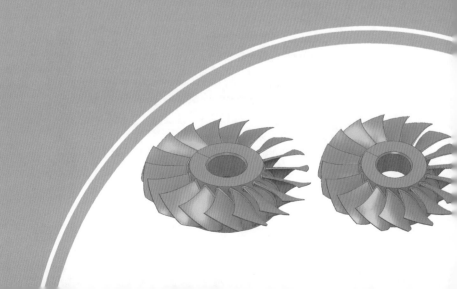

5.1 基本設定（Basic Set-up）

一、輸入模型（Import Model）

- 經由光碟 Chapter-05 輸入，Input model from CD Chapter-05: 3Plus2b.dgk

二、工作座標設定（Work Coordinate Setting）

- 框選全部模型 Select the model;
- 從物件管理列的工作座標，按滑鼠右鍵功能列點選「**建立和定位工作座標**」，Create and Orientate **Workplane**.
- 再從右鍵功能列點選「**物件頂部工作座標**」，Workplane at Top of Selection.

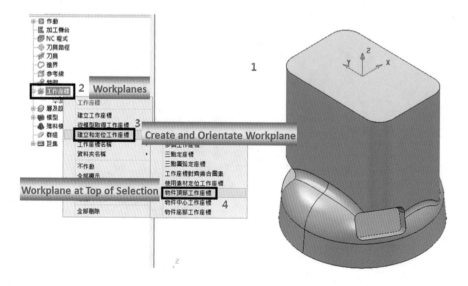

三、素材建立（Block Creation）

定義圓柱，選擇工作座標 1，計算，修改直徑為 180mm，接受

Under named coordinate, define cylindrical block and calculate, set diameter to 180mm.

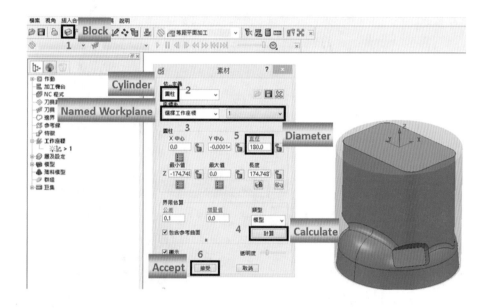

四、刀具設定以及刀具安全高設定（Setup for the tools, End-Mill for rough and Round-Nose for finish machining, Setup for safe-height to start cut）

- 從物件管理列的刀具，按滑鼠右鍵功能列點選「產生刀具」Create Tool
- 再從右鍵功能列點選「**圓鼻刀**」**Tip Radiused Tool**
- 從刀具選單中，定義直徑 Tool Diameter: 50mm, 圓角半徑 Radius: 5mm

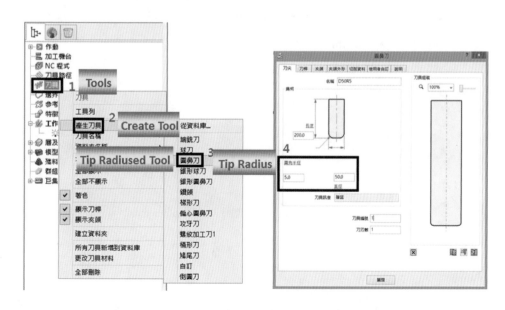

設定刀具安全高 Set safe toll height

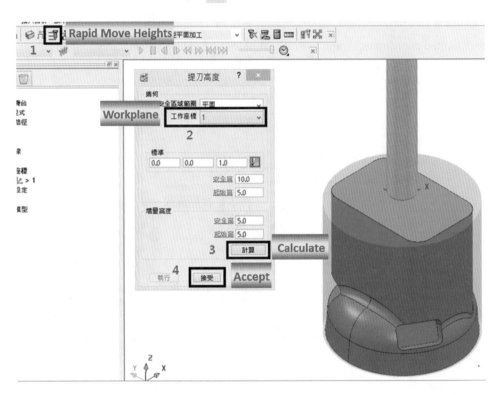

5.2　粗加工（Rough Machining in 3-Axis）

　　因模型的高度差使用大的刀具，所以粗加工可直接使用精加工工法的等高加工策略。

　　選擇工法選單 ⬧ 的等高加工工法，Because of the height, large tool is used in this example, therefore, "Constant Z Finishing" strategy can be used.

設定等高粗加工參數（Set parameter for the rough machining）

- 定義最小切削深度，define Stepdown
- 定義進退刀與連結，define leads and links

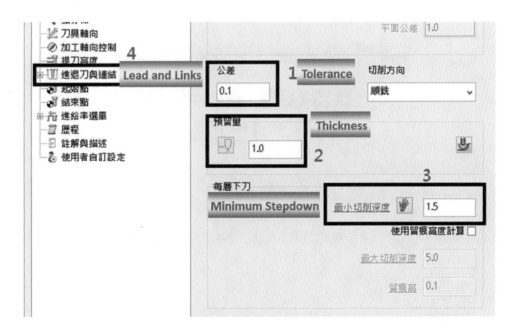

點選進退刀與連結，選擇曲面法線圓弧，定義角度 90°, 半徑 10, 進刀複製到退刀，Select leads and links option, define surface normal arc at 90°, radius at 10mm, Click to copy the Lead in values to the **Lead out**.

五軸銑削數控加工之基礎及實作

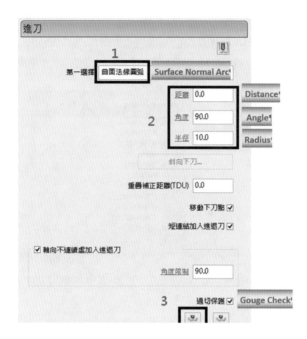

連結頁面中定義短連結用圓弧，長短連結分度值 30，Select links, define short of circular arc,set short/long threshold at 30.

設定限制條件，允許刀具軸線超出素材外面，Set limitation and allow tool axis outside the block.

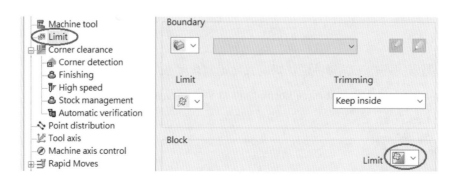

計算粗加工刀具路徑完成，Calculate and complete the tool path for rough machining as follow:

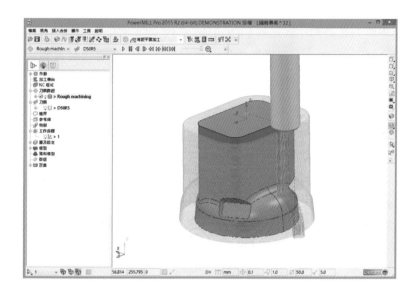

5.3　精加工（Finish Machining in 3-Axis）

同樣使用「等高精加工」工法和刀具，勾選螺旋狀功能，設定如下圖示的加工參數然後計算路徑 Select "Constant Z finishing" and the same tool, click "Spiral" option , Set up parameters and start to calculate.

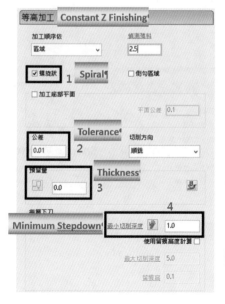

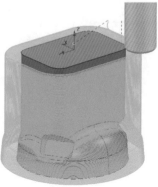

5.4 3+2 定面加工（Machining for the pockets with 3+2 axis）

定面加工及先將刀具軸向相對地轉到所要加工的平面，然後以 3- 軸的方式進行切削，因此也常常稱之爲 3+2 定面加工。刀具和工件的相對轉向則根據機台形式的不同而相異。For large portion of 5-Axis milling operation, the five axes are not move simutaneusly. In these cases, the tool-axis is first relatively rotate to the surface that to be machined, and then the tool remives the material in 3-Axis manner. This is also called 5-Face milling or 3+2 milling.

主要步驟如下，The procedue is as the following：

1. 產生局部工作座標，Create the work coordinate for pocket #1;
2. 作動局部工作座標，Activate the work coordinate;
3. 根據所選曲面產生局部素材，Calculate a block from the selected surfaces under activated WC;
4. 設定刀具安全高，Set safe tool height;
5. 建立殘料模型，Calculate a stock model;
6. 計算刀具路徑，Machining for the pocket;
7. 重復以上步驟加工其餘的口袋，Repeat steps 1-6 to machine other pockets。

具體圖解如下，Following are the pictorial explanations：

1. 爲加工口袋 #1 產生工作座標，Create the work coordinate for machining of pocket #1.
2. 作動此工作座標，Activate The work coordinate.

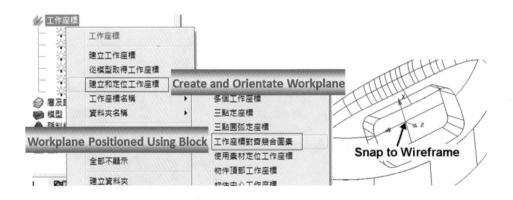

3. 以所選擇的曲面建立素材，Calculate a block from the selected surfaces

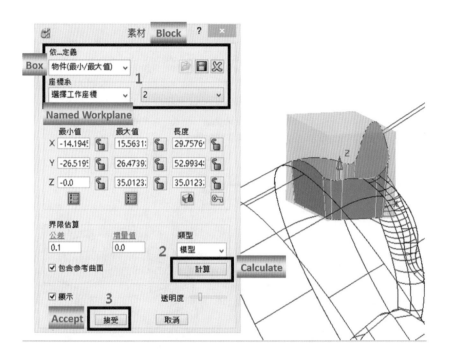

4. 設定刀具安全提刀高度，Set safe tool height

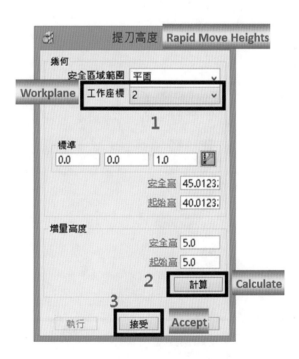

5. 建立殘料模型 - 刀間距定義 0.5，接受，Create the Stock Model-**Set up stepover at 0.5mm**.

將粗加工和精加工路徑新增到殘料模型做運算，All of toolpath added to stock model and calculate stock model.

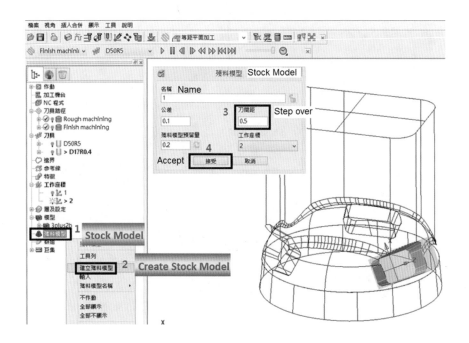

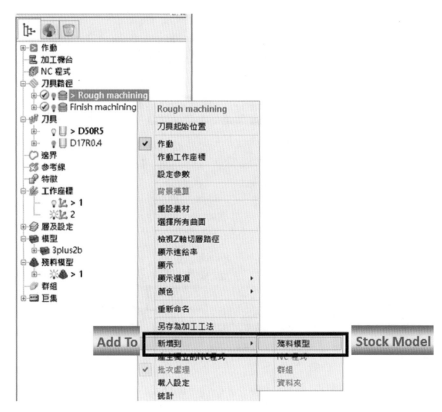

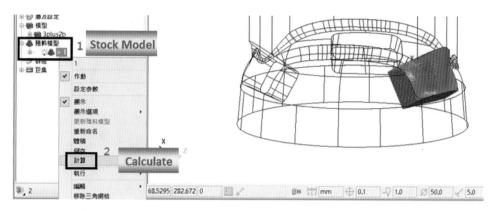

6. 刀具路徑工法選單 ：選擇模型餘料粗加工，Select "model rest area clearance" for rough machining;

- 從刀具選單中，定義圓鼻刀具直徑 17mm，圓角半徑 0.4mm，

  Select a Tip Radiused Tool, 17mm diameter and 0.4mm tip radius.

- 定義模型餘料加工參數，刀間距 12mm，每層下刀 1mm, Define parameters for the "Model Rest Area Clearance, Step over 12mm and Step down 1mm.

  定義參考殘料模型名稱，Define the name of this tool path.

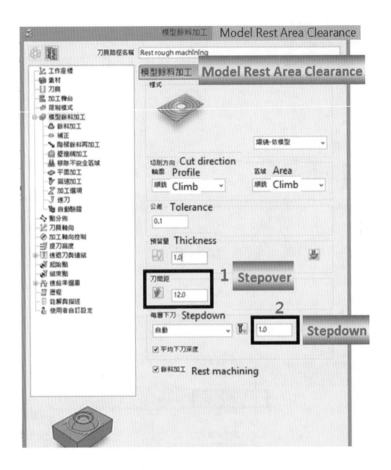

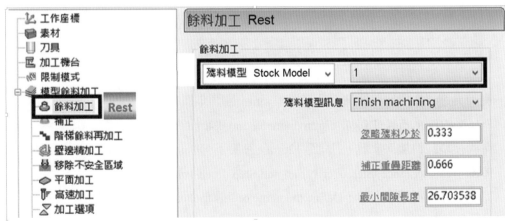

- 點選進退刀與連結，選擇進刀為斜向下刀，Click Lead/Link, select Lead.
- 安全高度的開始和終止點可定義為第一點安全高，The start and end safe height.
- 計算完成如下圖的口袋粗加工，Calculate to get the tool path as follow:

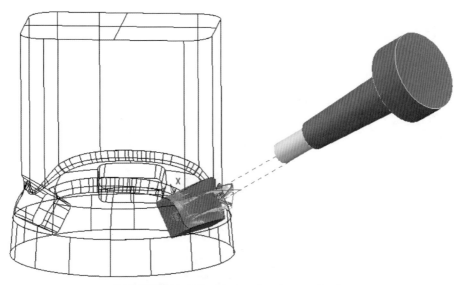

口袋的加工（Machining for the pocket）

加工其他兩個口袋時，必須分別以同樣方法建立相應的工作座標，素材以及殘料模型。

• 觀念導引 - Important concept introduction:

刀具路徑是以模型的加工工作座標系（名稱 1）作輸出，而非做路徑旋轉後的工作座標（名稱 2）。在自動轉換刀具軸向開啟狀態下，NC Data 會自動輸出旋轉的加工角度。

The NC program output is under the work coordinate WC1, not the work coordinate after rotating WC2. Under automatic tool axis transformation, NC program will automatically calculate the angle difference.

5.5 加工模擬（Machining simulation）

以第三章所述同樣的方法進行加工模擬，可以觀察加工模擬的狀況。You can simulate the machining in the same manner illustrated in Chapter 4.

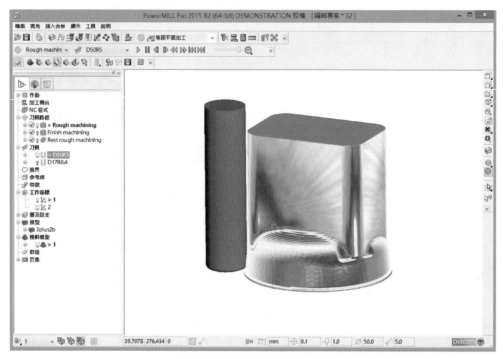

粗加工模擬（Simulation for rough machining）

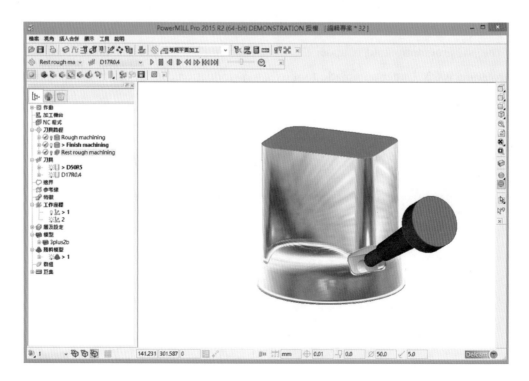

口袋加工模擬（Simulation for machining pocket-1）

五軸加工範例：哆拉 A 夢
（5-Axis Machining: Model of Doraemon）

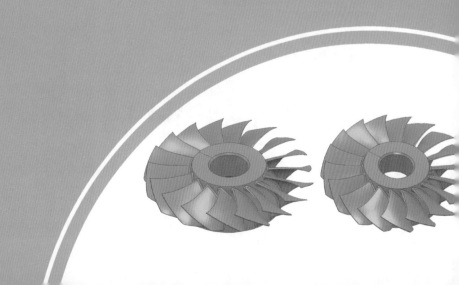

6.1 基本設定（Basic Setup）

一、輸入模型 Import Model：Doraemon.dgk（哆啦 A 夢）

經由光碟 Chapter-06 輸入：Doraemon.dgk, Import model from CD Chapter-06: Doraemon.dgk

二、素材定義（Create Block Material）

使用圓柱素材選項，高度最大值加至 0mm、最小值修改至 -100mm，直徑設定為 Ø65mm. At the selected work coordinate "Top", create a cylindrical block material, Z maximum at 0mm, Z minimum at -100mm, diameter of 65mm.

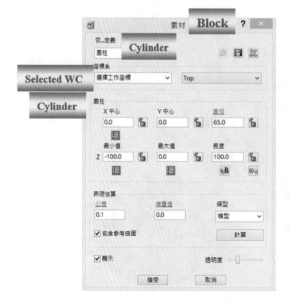

三、工作座標 Setup Work Coordinate

　　由於前半部的嘴巴部分有陡峭區，利用工作座標傾斜，以利使用粗加工。複製工作座標 "F" →重新命名爲 "Front", 並繞 Y 軸旋轉－15 度。要確認作動座標爲 "Front". Copy WC "F" and give it new name "Front". Rotate around y-axis at -15°. Make sure WC "Front" is activated.

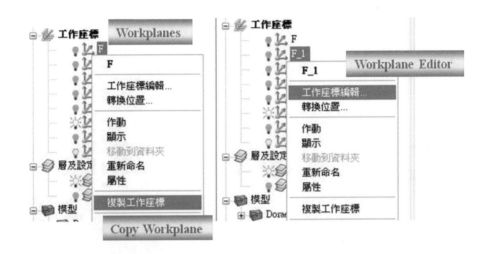

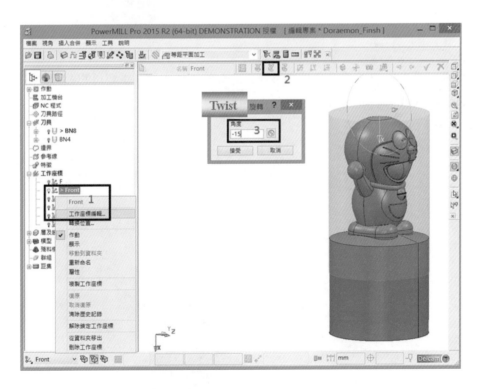

五軸銑削數控加工之基礎及實作

　　背面的座標做法同上步驟：複製工作座標 "B" →重新命名為 "Back" 並繞 Y 軸旋轉 15 度，（要確認作動座標為 "Back"）. Do the same thing for WC "B", rename to "Back" and rotate around y-Axis 15°. Make sure WC "Back" is activated.

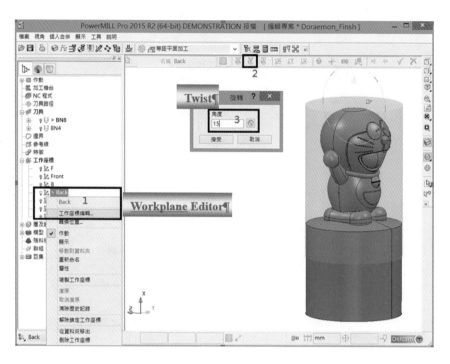

　　依模型的左、右、前、後分別建立不同 Z 方向的工作座標（主要目的做清角精加工）。
For clean corner machining, establish "L", "R", "F" and "B" for different z directions.

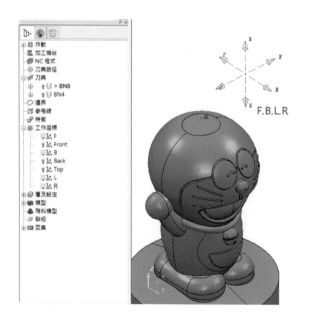

6.2　粗加工（Rough Machining）

一、粗銑（Rough Machining）

- 刀具設定 Tool setup：球刀直徑 Ball nosed mill of 8mm diameter，名稱 Name: BN8,

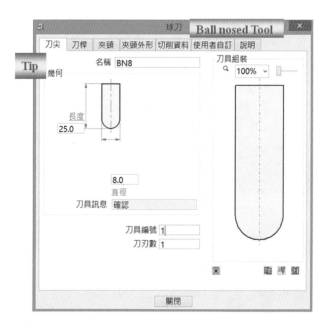

- 刀桿設定 Setup for shank：

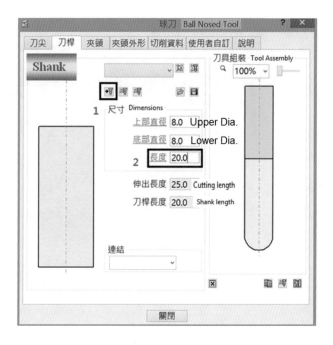

• 夾頭設定 Setup for tool holder：

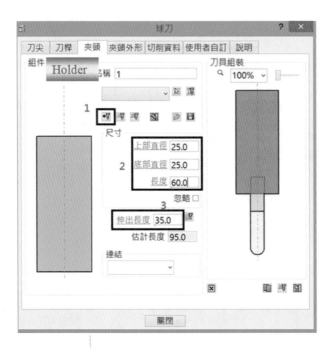

• 工法選單選擇 模型粗加工 Select "Model Area Clearance"；

　路徑名稱 Tool path name: Rou_Front；

• 選擇作動工作座標，Select Work coordinate -(Front)

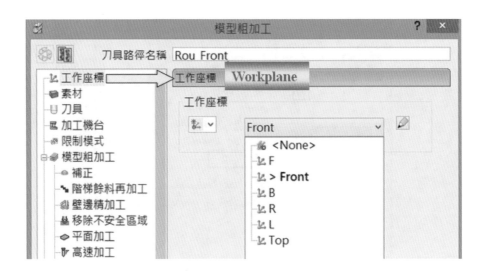

- 限制模式 - 最小值 Set Z Limit minimum -21mm,

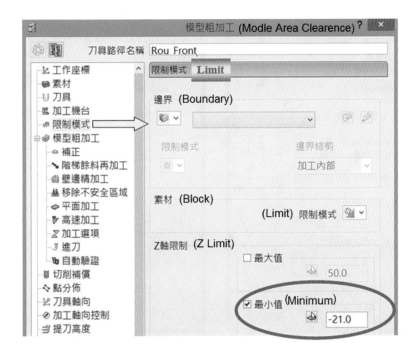

- 定義提刀高度，Define Rapid Move Heights

- 定義開始和終止點 ，Setup for Start and End point

- 定義進退刀連結 ，Setup for lead and links

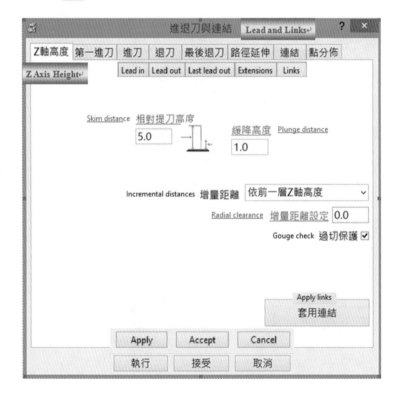

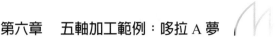

• 定義進刀爲斜向下刀，Setup Lead in for Ramp

• 確認連結的選項，Confirm Links

• 經由不等預留的選項功能定義干涉的曲面 ，Define interference surfaces. 定義干涉曲面的用意在於避免加工到此區域（下圖箭頭所指）所產生的刀具路徑，The purpose of the interference surfaces is to avoid machine the unnecessary area (As shown in the arrow pointed areas).

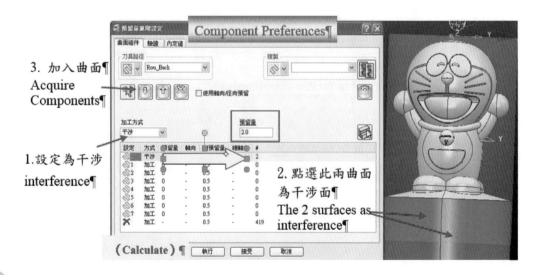

• 執行模型粗加工的工法，計算後路徑如下，Calculate to obtain the tool path:

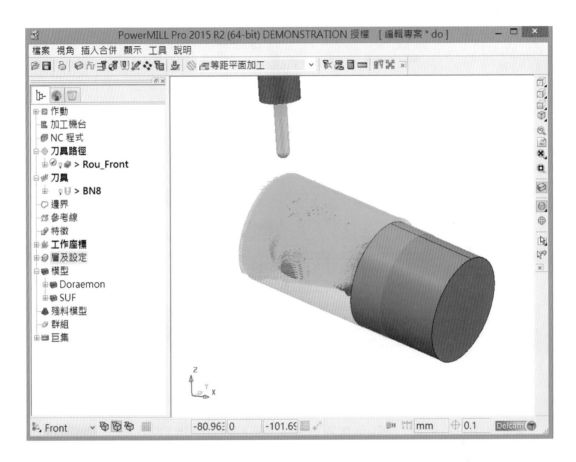

• 執行刀具路徑驗證 Verify the tool path just created：

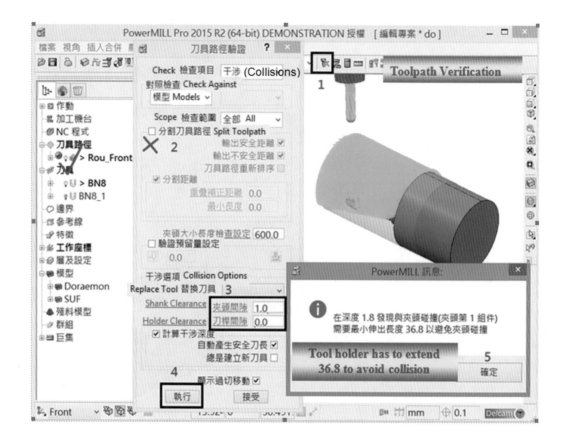

二、殘料粗加工（背面）Model rest area clearance (Back side)

- 刀具同樣設定 Tool setup：**Ball nose mill 球刀 R4**，名稱 Name :BN8

- 建立殘料模型 Create stock model：，並將 **Rou_Front** 路徑新增至殘

  料模型 Add tool path"Rou_Front" into the stock model，得到以下結果（建議殘料模型網

  格長度，Suggested step over : 0.2mm）

- 工法選單選擇 模型餘料加工，Select "**Model Rest Area clearance**";
  路徑名稱 Tool path name: **Rou_Back**;
- 選擇作動工作座標 Select Work coordinate-(**Back**);
- 定義提刀高度（切換 Back 重新運算安全高），Re-Calculate Rapid Move Heights;
- 不等預留量選項中，複製「**Rou_Front**」路徑的干涉曲面 > ，In "Default Thickness Option" , "Copy Thickness Data" .
- 其餘的選項參數定義不變同「**Rou_Front**」路徑，Other parameters are the same as in tool-path "**Rou_Front**".

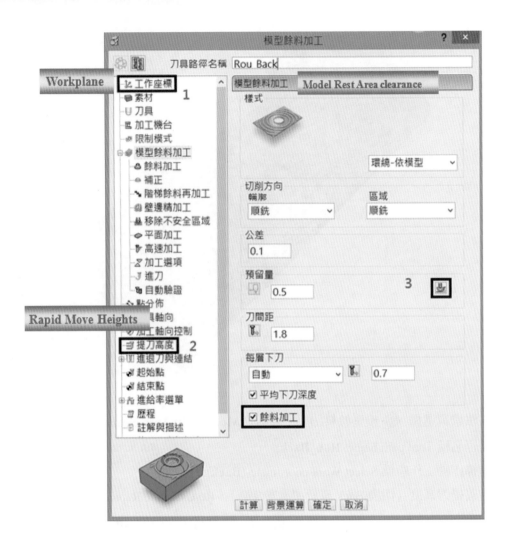

• 選擇參考的殘料模型，Select reference stock model

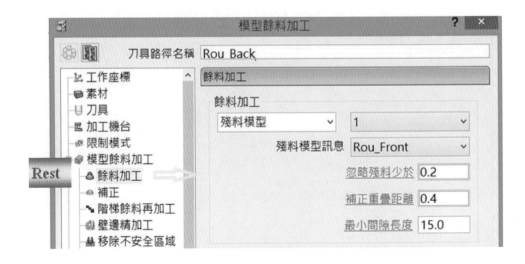

• 執行模型餘料加工的工法，計算後路徑如下，Calculate to obtain the tool path:

• 執行刀具路徑驗證 Toolpath verification：

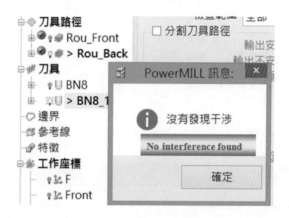

• 將 Rou_Back 路徑加入殘料模型運算，結果如下圖示，add the "Rou_Back" tool path into stock model and calculate and obtain the following:

三、沿面投影 - 中銑（Semi-finish: Surface Projection Finishing）

- 模型右鍵功能中點選輸入參考曲面，Right click "Model" and select "Import Reference Surfaces"

- 經由光碟 Chapter-06 輸入參考面，Import reference surface from CD Chapter-06: **SUF.dgk**

- 工法選單選擇 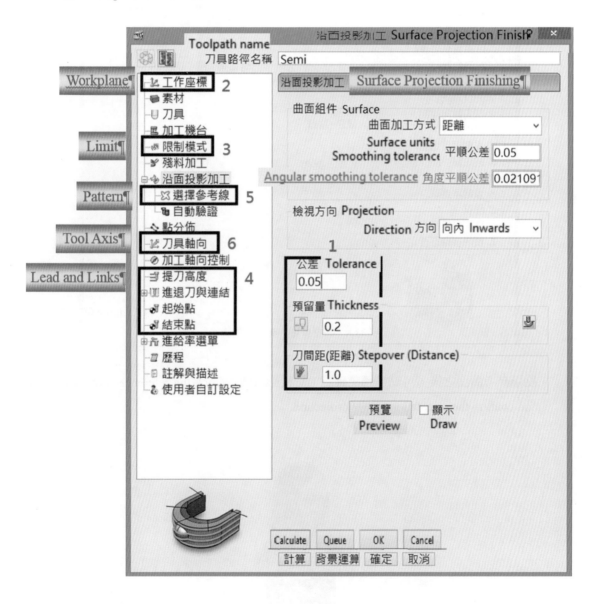 沿面投影加工，Select "Surface Projection Finishing"
- 路徑名稱 Tool path **Name**: Semi
- 選擇作動工作座標，Select Work coordinate (Top);
- 限制模式 - 最小值，Setup Limit, minimum -80;
- 定義提刀高度（切換 Top 重新運算安全高），Re-Calculate Rapid Move Heights;
- 定義進退刀與連結，進退刀可選擇曲面法向圓弧／角度／半徑，短連結可選擇圓弧連結，Setup Lead and links.

- 定義參考線投影的 UV 方向選擇，選擇螺旋狀功能，Define "Pattern" and select "Spiral".

- 點選參考曲面之後，可點選預覽確認 UV 投影方向，Click the reference surface, preview and confirm the orientation of UV projection.

- 定義刀具軸向，Setup Tool-Axis

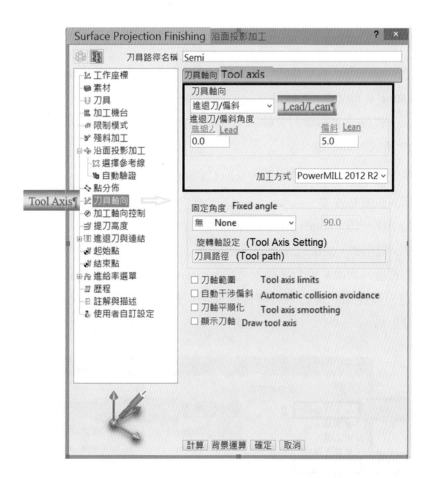

- 執行沿面投影中銑加工，計算後路徑如下，Calculate to obtain the tool path：

6.3 精加工（Finish Machining）

一、清角精加工：Corner finishing

分別針對前、後、左、右來作定軸清角精加工，路徑名稱 **Tool path name: Rest_F, Rest_B, Rest_L and Rest_R** for front, back, left and right side respectively.

- 刀具設定：直徑為 4 的球刀並輸入名稱 BN4，Tool setup: Create ball mill of 4mm in diameter，Tool Name: BN4
- 工法選單選擇 ▨ 自動清角加工，Select "Corner finishing"
- 路徑名稱 Name: Rest_F
- 選擇作動工作座標，Select Work coordinate - (F);
- 限制模式 - 最小值，Setup Limit, minimun -2;
- 定義提刀高度（切換 F 重新運算安全高），Re-Calculate Rapid Move Heights;
- 定義進退刀與連結，進退刀可選擇曲面法向圓弧，短連結可選擇圓弧連結，Define leads and links

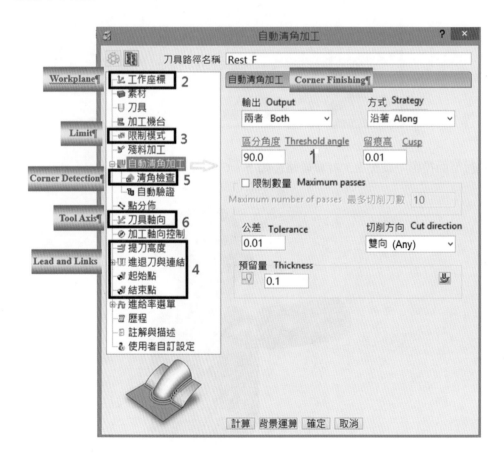

- 清角檢查選項中定義參考刀具 BN4 與重疊補正距離 0.2mm，Run Corner Detection for Tool BN4 and Overlap as 0.2mm.

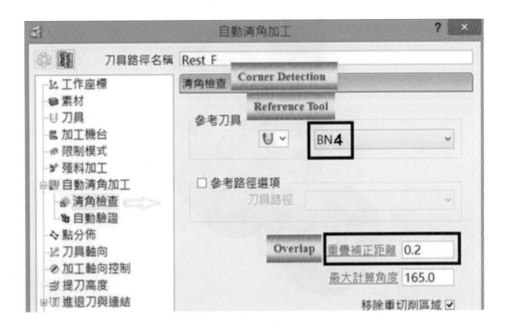

- 刀具軸向定義垂直軸向，Define Tool Axis Orientation as "Vertical".

• 執行清角加工，計算後路徑如下，Run "Corner Finishing", calculate to obtain the tool path:

• 後側、左側、右側清角加工路徑請參考上述作法 Repeat above procedures for back, left and right side toolpaths, Names：Rest_B、Rest_R、Rest_L.

• 需要切換選擇不同的作動工作座標 - (B/R/L)，Need to activate Work coordinate;

• 需要切換 B/R/L 重新運算安全高度，Need to re-calculate Rapid Move Heights; for all activated work coordinates (**B/R/L**);

此模型你可選擇只做清角加工
正反（Rest F & Rest B）兩面即可
For this model, only Rest F
and Rest B need to be Corner-Finish.

二、細銑加工（Finishing）

操作同中銑加工的沿面投影加工策略，你可複製 Semi 路徑更改切削參數條件做細銑加工，You can copy the "Semi-Finish" toolpath and modify parameters to do finishing job. 或者重複以下的操作，Or repeat the following procedure:

- 工法選單選擇 沿面投影加工，Select toolpath strategy "**Surface Projection Finishing**"
- 路徑名稱，Tool path Name: **Finish**
- 選擇作動工作座標，Select Work coordinate -(**Top**);
- 限制模式 - 最小值，Setup Limit, Minimum -80;
- 定義提刀高度（切換 Top 重新運算安全高），Re-Calculate Rapid Move Heights;
- 定義進退刀與連結，進退刀可選擇曲面法向圓弧，短連結可選擇圓弧連結，Setup Leads/Links,

注意：**需選取參考曲面**（Note: Select the reference surface）SUF.dgk,

- 執行沿面投影細銑加工，計算後路徑如下，Calculate to obtain the tool path:

執行刀具路徑驗證，Execute toolpath verification [圖] .

所有路徑計算完後，務必執行干涉檢查，確保路徑安全性，

All completed toolpath have to conduct collision-checking to make certain they are safe to run.

7

五軸加工範例：足球
（5-Axis Machining: Soccer Ball）

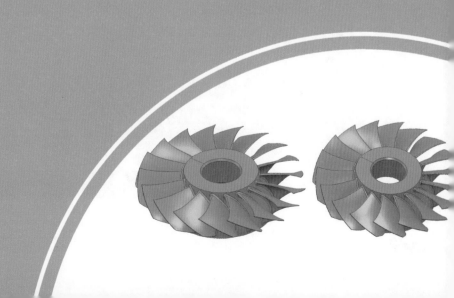

7.1 基本設定（Basic Setup）

一、輸入模型，mport Model: Soccer ball（足球）

經由光碟 Chapter-07 輸入模型，Import from CD Chapter-07: **Soccer ball.dgk**

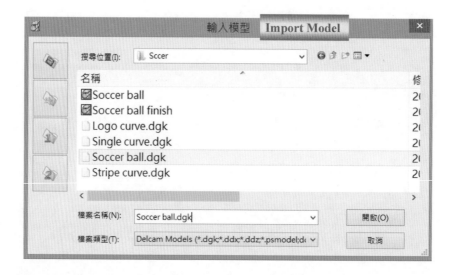

二、素材定義 Setup Block Material

使用圓柱素材選項，高度最大值加大 1.2mm，直徑設定為 ø60mm, **Setup cylindrical material 60mm in diameter and add 1.2mm in to the height**.

三、工作座標 Work Coordinate

依照足球的前、後分別建立不同 Z 方向的工作座標（Y 軸方向一致），Establish four work coordinates, with y in the same orientation.

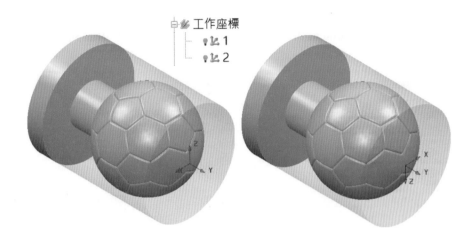

四、建立刀具 Create Tool

- 刀具設定：

    1. **圓鼻刀** 直徑 D10 鼻端圓角半徑 R1；名稱：D10R1,

       Create a round nosed tool with diameter of 10mm, round edge of 1mm.

    2. 球刀，直徑 8mm，名稱：BN8

       Create a ball mill tool of 8mm diameter, name BN8

    3. 球刀，直徑 2mm，名稱：BN2

       Create a ball mill tool of 2mm diameter, name BN2

    4. 球刀，直徑 1mm，名稱：BN1

       Create a ball mill tool of 1mm diameter, name BN1

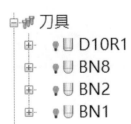

7.2 粗加工（Rough Machining）

一、粗銑（Rough Machining）

工法選單選擇 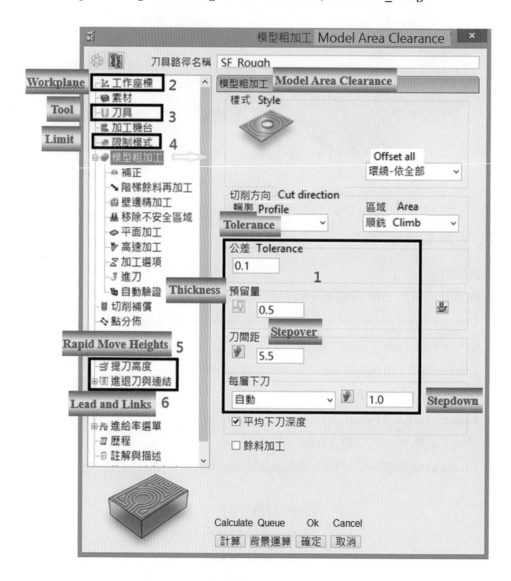 模型粗加工，路徑名稱 SF_Rough;

Create tool path: Rough machining for left-hand side, Name: SF_Rough

- 選擇作動工作座標，Select Work coordinate -(1);
- 選擇圓鼻刀 Select Tip Radiused tool: D10R1;
- 限制模式最小值 Select Limit minimum: -2mm ,

- 定義提刀高度（切換 1 座標重新運算安全高），Re-calculate fine Rapid Move Heights;
- 定義進退刀連結 ，Setup for leads and links

- 定義進刀為斜向下刀，Setup Lead in for Ramp option:

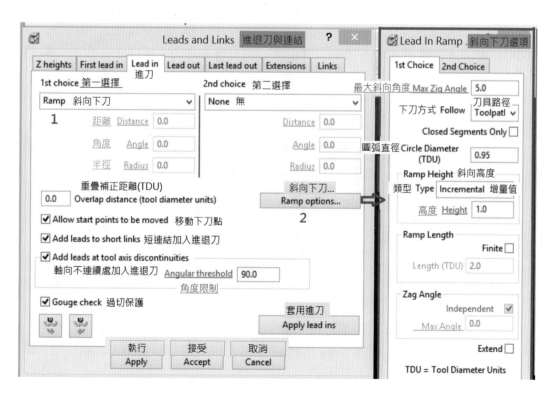

• 定義開始和終止點 ，Setup for Start and End point

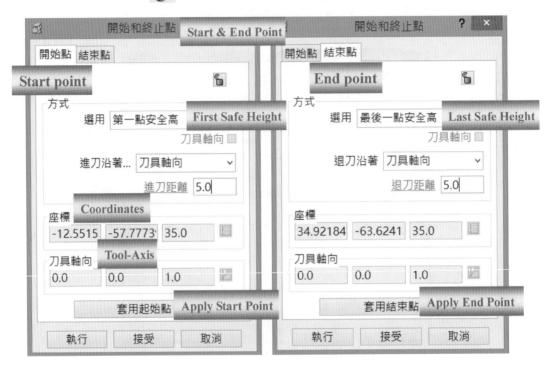

• 執行模型粗加工的工法，前側計算後路徑如下，Calculate to obtain the tool path:

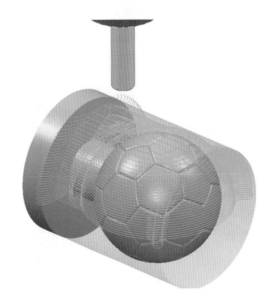

後側粗加工：SB_Rough 路徑，請參考上述作法或複製 SF_Rough 路徑，Rough machining for the right-hand side is in same manner or copy SF_Rough toolpath.

- 需要切換選擇不同的作動工作座標，Need to change Work coordinate-(2);
- 需要切換 (2) 重新運算安全高度，Re-calculate Rapid Move Heights;
  計算後路徑如下，Calculate to obtain the tool path:

二、中銑（Semi-Finish）

- 工法選單選擇 點投影加工，Select "Point Projection Finishing";
- 路徑名稱 ，Tool Path Name: Semi;
- 無需作動任何工作座標（即是使用世界座標），Don't select any Work coordinate;
- 選擇球刀，Select Ball Nose tool BN8;
- 限制模式，最小值，Setup Limit minimum -60mm.

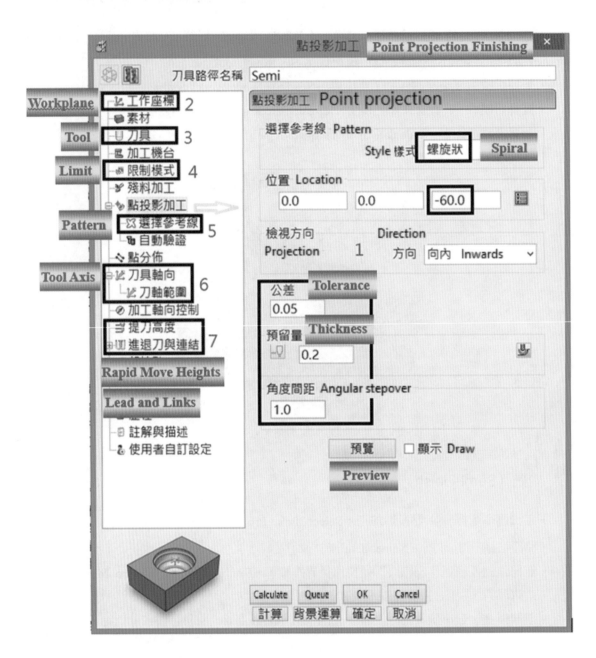

- 選擇參考線投影的限制範圍，選擇螺旋狀功能，Select Limits in "Pattern", and "Spiral"；
  點選預覽確認投影的位置與範圍是否正確，Preview to confirm.

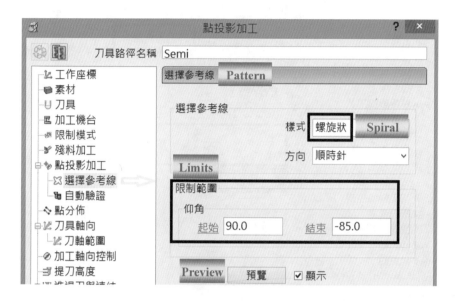

- 定義刀具軸向，Setup Tool-Axis;

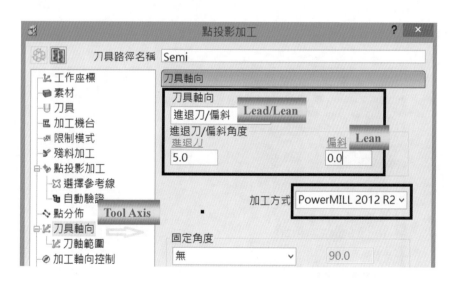

- 定義刀具軸向限制，Setup Tool-Axis limits;

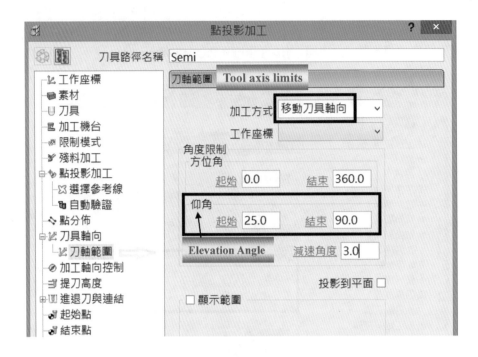

- 定義提刀高度 ▆ （切換 None 重新運算安全高），Re-Calculate Rapid Move Heights;
- 定義進退刀與連結 ▆ ，進退刀可選擇垂直圓弧／角度／半徑，短連結可選擇圓弧連結，Setup Leads and Links.

- 執行點投影中加工的工法，計算後路徑如下，Calculate to obtain the tool path:

三、細銑：點投影加工 Point Projection Finishing

複製中銑 Semi 路徑，Copy "Semi" toolpath

複製 Semi 路徑只需更改以下選項，Copy "Semi" tool path procedure:

• 路徑名稱 Toolpath Name: Finish

• 切削參數條件 Set cutting parameters

• 選擇同樣球刀 Select same Ball Nosed tool BN8;

注意到：與中銑 "Semi" 不同之處爲公差＝ 0.01；預留量＝ 0.0. Note: The only differences
　　　　are the Tolerance=0.01 and Thickness=0.0.

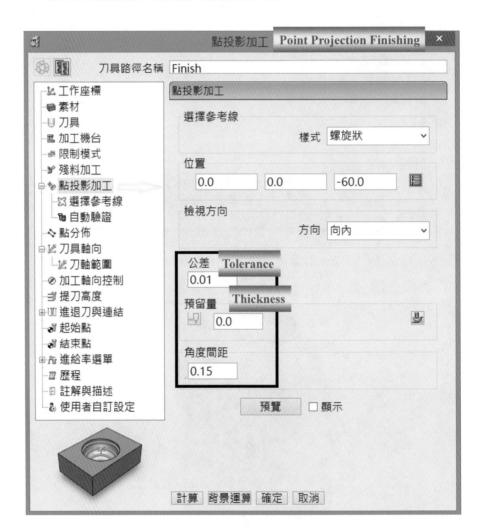

• 執行點投影細加工的工法，計算後路徑如下，Calculate to obtain the tool path:

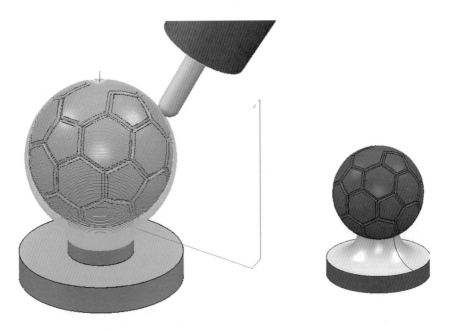

四、底座細洗 Finish for the Base

- 經由光碟 Chapter-07 輸入參考面：載入後並選取參考面

  From CD chapter-07 Import and select reference surface: surface.dgk

- 工法選單選擇 沿面投影加工 Select "Surface Projection Finishing"
- 路徑名稱 Toolpath Name: Base
- 選擇同樣球刀 Select Ball Nosed tool BN8;
- 限制模式 - 最大值 Setup Limit, Maximum -60;

五軸銑削數控加工之基礎及實作

• 定義進退刀與連結，進退刀可選擇曲面法向圓弧／角度／半徑，短連結可選擇圓弧連結，Define Lead and Links;

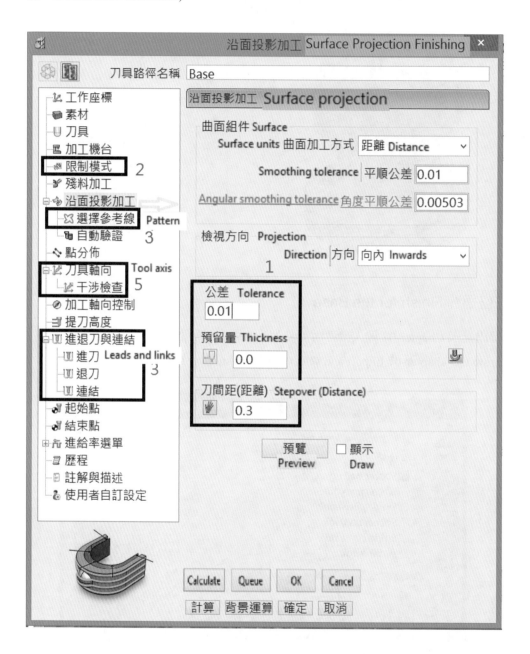

• 定義參考線投影的 UV 方向選擇，選擇螺旋狀功能，Define Pattern direction as U and check **Spiral**，點選參考曲面之後，可點選預覽確認 UV 投影方向，After click the reference surface, preview and confirm.

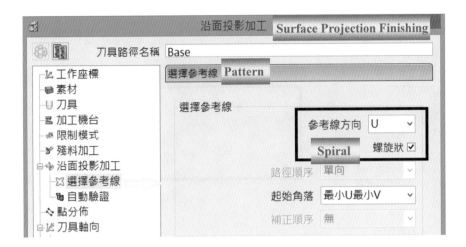

- 定義刀具軸向，Setup Tool-Axis
- 勾選自動干涉偏斜，Select automatic collision avoidance

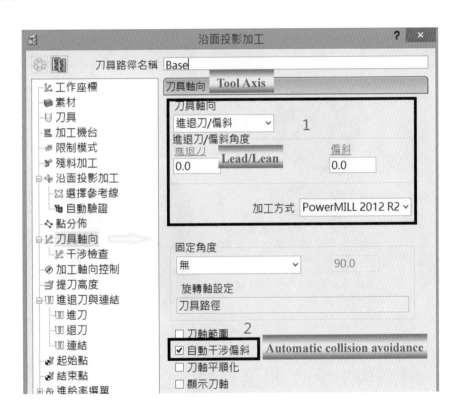

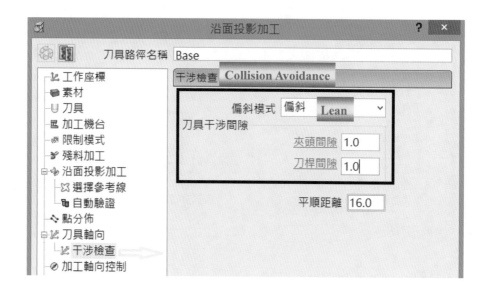

- 執行沿面投影底座加工的工法，計算後路徑如下，Calculate to obtain the tool path:

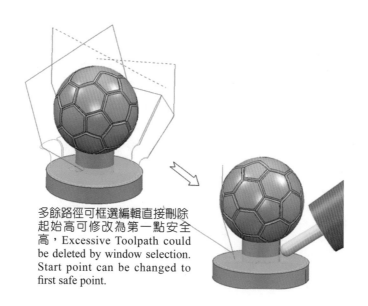

多餘路徑可框選編輯直接刪除
起始高可修改為第一點安全
高，Excessive Toolpath could
be deleted by window selection.
Start point can be changed to
first safe point.

7.3　清溝與 Logo 加工（Machining Logo）

一、單刀清溝：參考線投影加工 Pattern (Single Curve)

- 模型右鍵功能中點選輸入參考曲線，Right click the model, select patterns

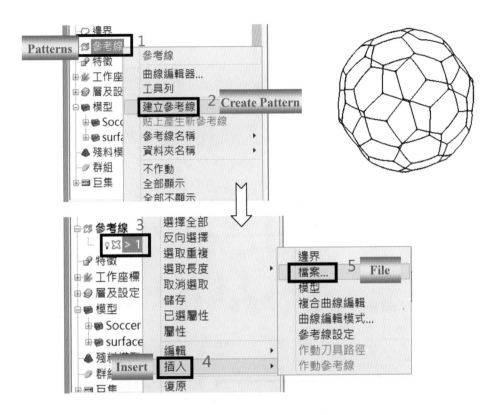

- 經由光碟 Chapter-07 輸入參考曲線（更名為 Single curve）：
  Import and select pattern (reference curve): "Single curve.dgk"

- 工法選單選擇 參考線投影加工 Select Tool Path: "Pattern Finishing"
- 路徑名稱 Tool Path Name: "Single curve";
- 選擇球刀 Select ball nosed tool BN2;

五軸銑削數控加工之基礎及實作

• 限制模式 - 最小值 Setup Limit, Minimum: -51mm.

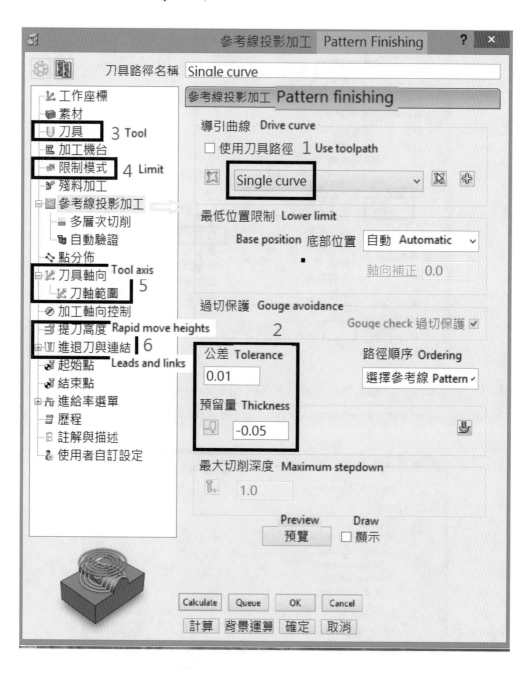

128

- 定義刀具軸向，Setup Tool-Axis

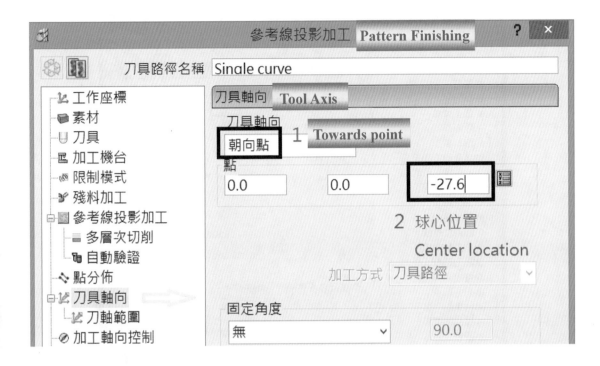

- 定義刀具軸向限制，Setup Tool-Axis limits

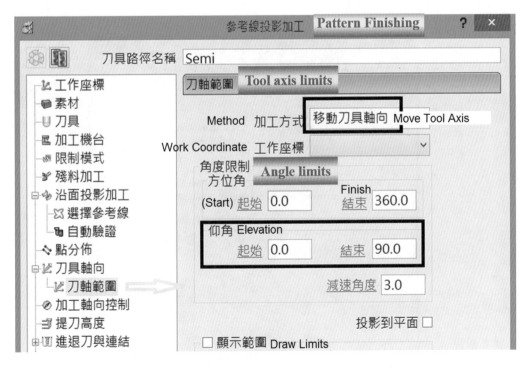

• 定義提刀高度〔切換（球 Sphere）重新運算安全高〕，Define Rapid Move Heights;

• 定義進退刀連結 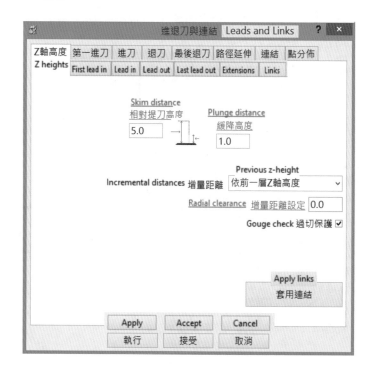，Setup for leads and links

• 定義進退刀為曲面法線圓弧，Setup Lead in/out for Surface normal arc

• 定義連結為圓弧，Setup Link for Circular arc

• 執行參考線投影清溝加工的工法，計算後路徑如下，Calculate to obtain the tool path:

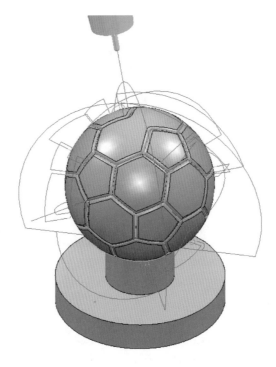

你可透過路徑編輯工具列
自動調整加工順序優化
You can re-order toolpath segment

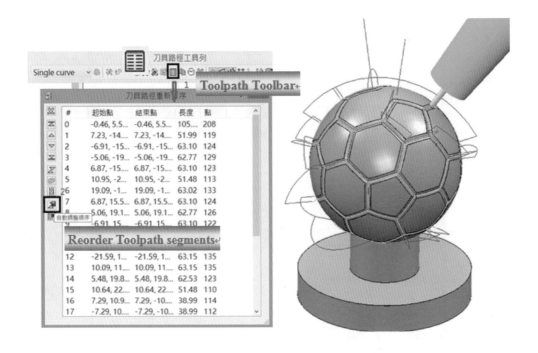

#	起始點	結束點	長度	點
0	-0.46, 5.5....	-0.46, 5.5....	105....	208
1	7.23, -14....	7.23, -14....	51.99	119
2	-6.91, -15....	-6.91, -15....	63.10	124
3	-5.06, -19....	-5.06, -19....	62.77	129
4	6.87, -15....	6.87, -15....	63.10	123
5	10.95, -2....	10.95, -2....	51.48	113
26	19.09, -1....	19.09, -1....	63.02	133
7	6.87, 15.5....	6.87, 15.5....	63.10	124
8	-5.06, 19.1....	5.06, 19.1....	62.77	126
9	-6.91, 15	-6.91, 15	63.10	122
12	-21.59, 1....	-21.59, 1....	63.15	135
13	10.09, 11....	10.09, 11....	63.15	135
14	5.48, 19.8....	5.48, 19.8....	62.53	123
15	10.64, 22....	10.64, 22....	51.48	110
16	7.29, 10.9....	7.29, -10....	38.99	106
17	-7.29, 10....	-7.29, -10....	38.99	112

Single curve

刀具路徑工具列

Toolpath Toolbar

刀具路徑重新排序

自動調整順序

Reorder Toolpath segments

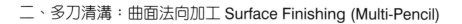

二、多刀清溝：曲面法向加工 Surface Finishing (Multi-Pencil)

- 工法選單選擇 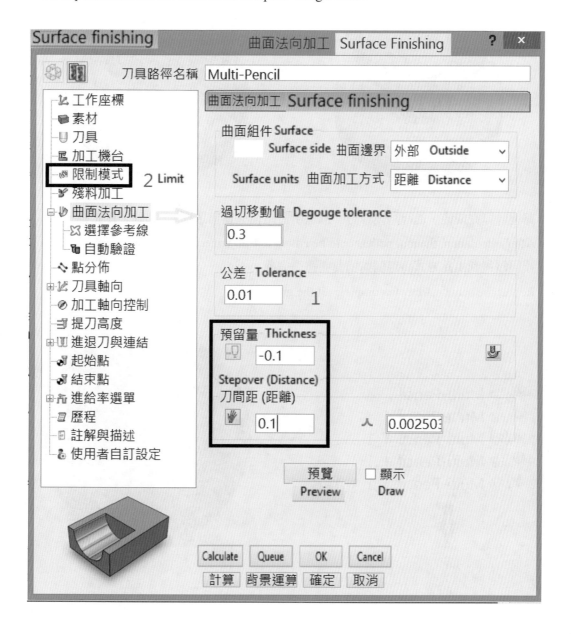 曲面法向投影加工，Select "Surface Finishing";
- 路徑名稱，Toolpath Name: Multi-Pencil;
- 同樣選擇球刀，Select Ball Nosed tool BN2;
- 確認取消限制模式，Cancel Limit;

其餘參數與 Single curve 刀具路徑相同，

Other parameters are the same as in Toolpath "Single curve"

• 執行曲面法向投影加工的工法，計算後路徑如下，Calculate to obtain the tool path:

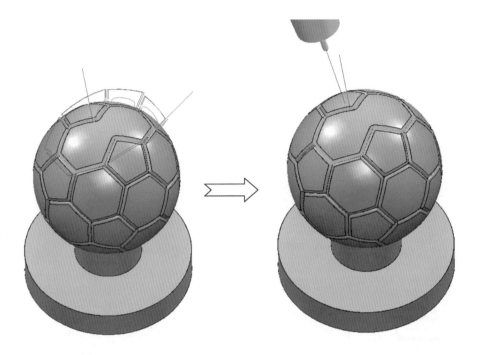

路徑計算完成後，進行路徑修剪（框選所需路徑 > 反向選擇），移除多餘路徑，

After calculation, the tool path should be edited to trim unnecessary paths.

複製路徑利用此方式將每個溝槽曲面路徑完成，

Repeat the same manner to complete all the grooves.

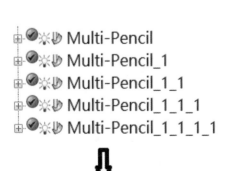

接下來進行路徑合併，按住 Ctrl 鍵 滑鼠點選要合併的路徑，拖移至主要路徑裡，Append tool paths by dragging intended tool path into the major tool path, with the Ctrl key pressed.

完成後路徑如下 The tool paths are as the following after finish:

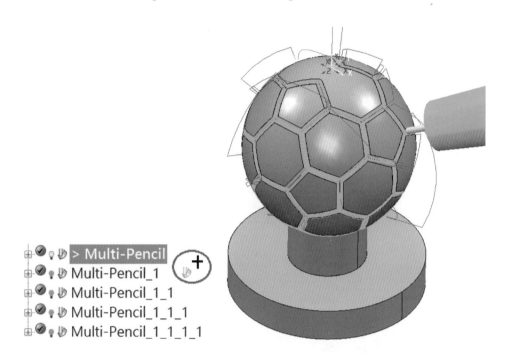

三、多邊形法線參考線加工，Embedded Pattern Finishing:

使用法線投影工法主要目的在於線段 Logo 文字可做五軸的法向角度加工，但需要將一般的參考線轉換成法線參考線（與曲面有 normal 關聯）才可選用，操作如下，The usage of the "Embedded Pattern Finishing" is to machine the letters of the Logo. But need to transform regular "reference curve" into "Embed". The procedure:

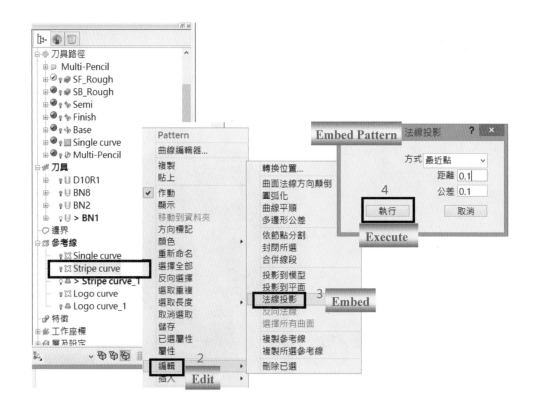

工法選單選擇 法線投影加工 Toolpath: "Embedded Pattern Finishing";

- 路徑名稱 **Toolpath Name**: "Stripe curve machining"
- 選用球刀名稱：Ball Nosed Tool, 1mm, Name: BN1

• 取消限制模式 - 最小值，**Cancel** Limit, Minimum -51;

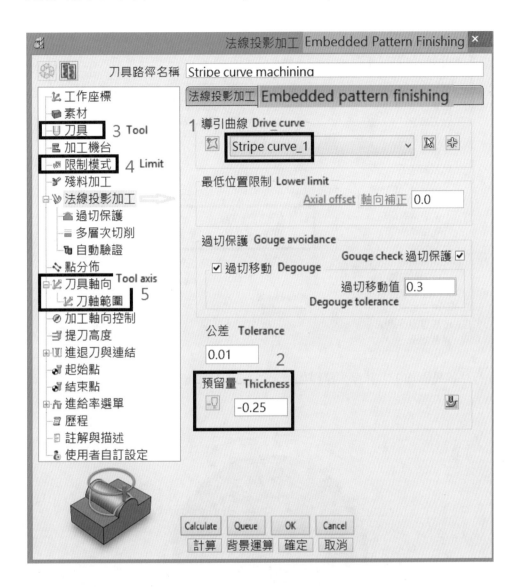

• 定義刀具軸向，Setup Tool-Axis

- 定義刀具軸向限制，Setup tool axis limits

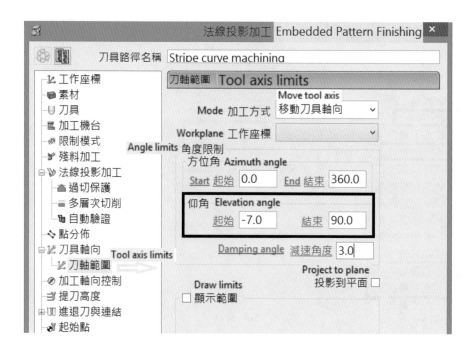

- 執行法線投影加工，計算後路徑如下，Calculate to obtain the tool path:

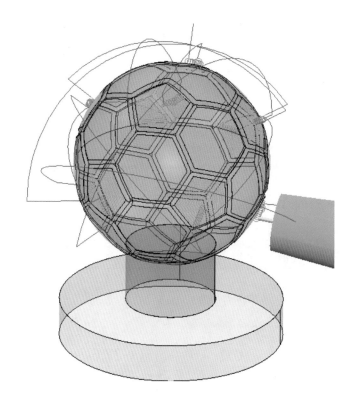

四、Logo 加工（Machining the Logo）

複製多邊形路徑，Copy toolpath "**Stripe curve machining**",

複製以上路徑只需更改以下選項，The procedure of copy "**Stripe curve machining**":

- 路徑名稱，**Name: "Logo machining"**;
- 選用 Logo curve_1 法線參考線，Select "**Logo curve_1**" as Drive curve;
- 選擇同樣球刀，Select same Ball Nosed Tool: **BN1**;
- 定義進退刀與連結，進退刀設定無（None），短連結選擇沿曲面連結（on Surface），
  Define Leads and Links: "None" for Lead-in & Lead-out; "on Surface" for Short link.

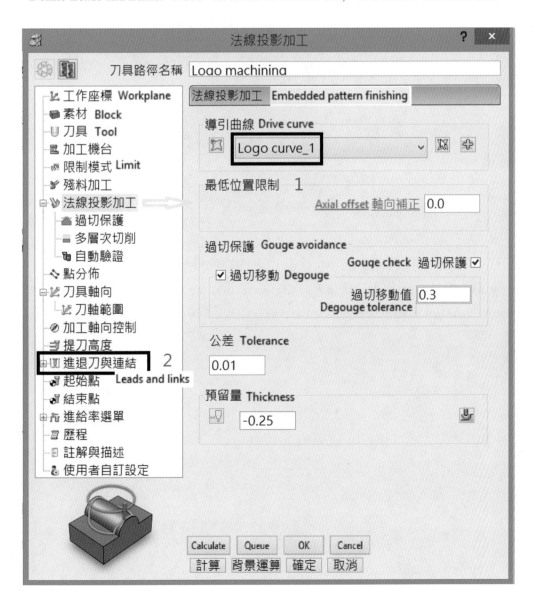

• 執行法線 Logo 投影加工，計算後路徑如下，Calculate to obtain the tool path:

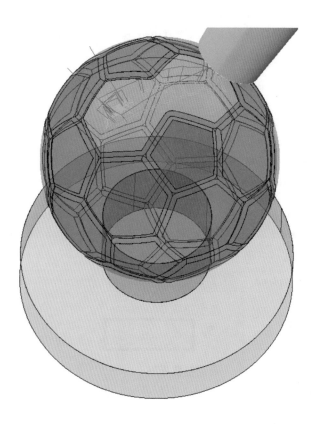

所有路徑計算完後，務必執行干涉檢查，確保路徑安全性

Now overall tool paths are completed. Before export the NC programs for machine center, interruption checkups have to be conducted to prevent any collisions in real machining process.

8

五軸加工範例：觀音像
（5-Axis Machining: Statue of
Guan-Yin, Goddess of Mercy）

8.1 基本設定（Basic Setup）

一、輸入模型：Mercy Buddha（觀音像）Import Model

經由光碟 Chapter-08 輸入模型。Import model from CD Chapter-08: **Mercy Buddha.dgk**

二、素材定義 Block Material Definition:

使用圓柱素材選項，高度最大值加至 100mm、最小值修改至 7.5mm，直徑設定為 Ø60mm, Create a cylindrical block of 100mm in height and 60mm in diameter, the minimum value of Z changed into 7.5mm.

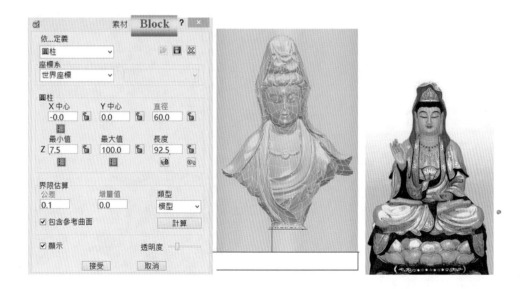

三、工作座標（Work Coordinate）

依照觀音像的左、右、前、後分別建立不同 Z 方向的工作座標（Y 軸方向一致），
Establish four work coordinates, with Y in the same orientation.

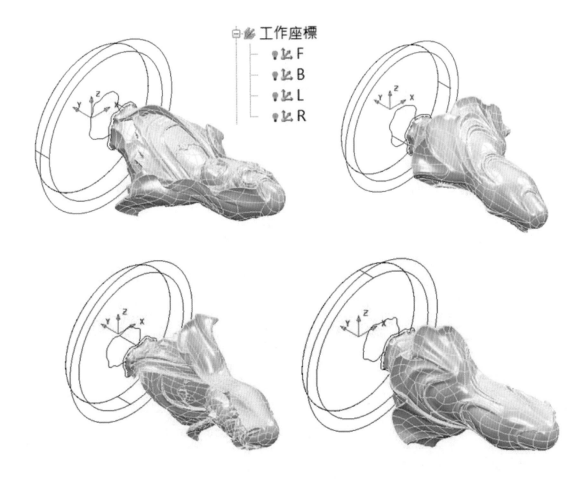

四、建立刀具（Create Tools）

- 刀具設定，Setup tools：

1. **圓鼻刀** 直徑 D10 鼻端圓角半徑 R1；名稱：D10R1,

   Create a round nosed tool with diameter of 10mm, round edge of 1mm, name: D10R1

2. 球刀，直徑 6mm，名稱：BN6

   Create a ball mill tool of 6mm diameter, name: BN6

3. 球刀，直徑 4mm，名稱：BN4

   Create a ball mill tool of 4mm diameter, name: BN4

4. 球刀，直徑 3mm，名稱：BN3

   Create a ball mill tool of 3mm diameter, name: BN3

5. 錐形圓鼻刀 4mm，名稱：D4_R

   Create a tapered tipped tool of 4mm diameter, name: D4_R

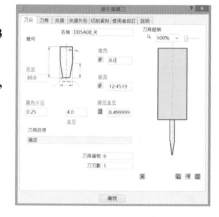

✎ 8.2 粗加工（Rough Machining）

一、粗銑（Rough Machining）

1. 工法選單選擇 ▨ 模型粗加工，路徑名稱 L_Rough;

   Create tool path: "Model Area Clearance" for left-hand side, Name: L_ROUGH

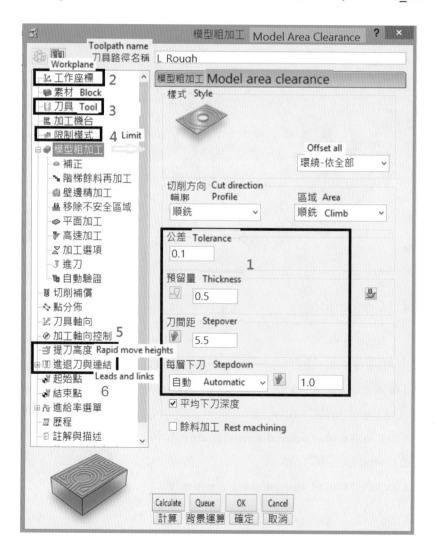

- 選擇作動工作座標，Select Work coordinate-(**L**);
- 選擇圓鼻刀，Select Tip Radiused tool **D10R1**;
- 限制模式 - 最小值，Select Limit, Minimum -5 ;
- 定義提刀高度（切換 L 座標重新運算安全高），Calculate Rapid Move Heights;
- 定義進退刀連結 ，Setup for leads and links;

- 定義進刀為斜向下刀，Setup Lead in for Ramp

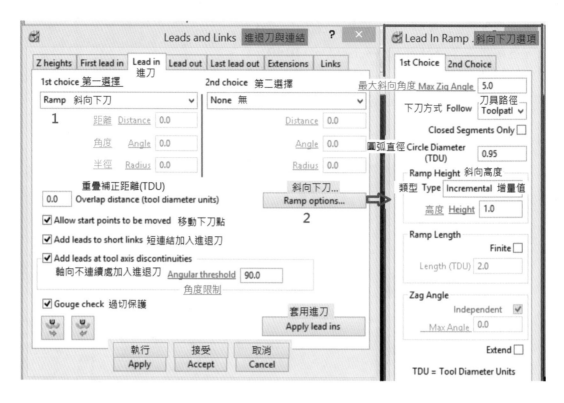

• 定義開始和終止點 ，Setup for Start and End point

• 執行模型粗加工的工法，計算後路徑如下，Calculate to obtain the tool path:

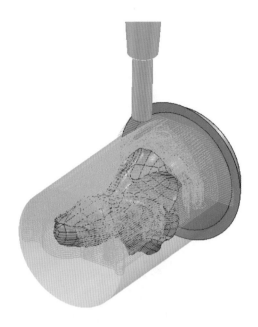

右側粗加工：R_ROUGH 路徑，請參考上述作法，Machining for the right-hand side is in similar manner.

• 需要切換選擇不同的作動工作座標，Change Work coordinate-(R);

• 需要切換（R）重新運算安全高度，Recaculate "Rapid Move Heights";

計算後路徑如下，Calculate to obtain the tool path:

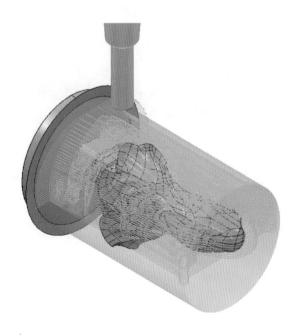

二、餘料加工 Rest Area Machining

建立殘料模型：，並將 L_ROUGH 與 L_ROUGH 路徑新增至殘料模型，得到以下結果（殘料模型網格長度：建議 0.2）。

Establish stock model: , add the above rough machining tool paths **L_ROUGH & L_ROUGH** into this stock model, calculate the stock model and find the result in the following picture. Step over of 0.2mm is suggested.

顯示選項 > 著色
Drawing options > Shaded

- 工法選單選擇  模型餘料加工，Select **"Model Rest Area clearance"**;
  路徑名稱，Tool path name: **Rest_1**;
- 同樣作動工作座標，Select Work coordinate **-(R)**;
- 選擇球刀，Select Ball Nosed tool **BN4**;
- 限制模式 - 最小值，Select Limit - Minimum - 0 ,

- 選擇參考的殘料模型，Select reference stock model

- 定義提刀高度（R 工作座標運算安全高），Calculate Rapid Move Heights under WC(R);
- 經由不等預留的選項功能定義干涉的曲面 <img>，Define interference surfaces,
  定義干涉曲面的用意在於避免加工到此區域所產生的刀具路徑，This is to avoid
  machine un-necessary areas.

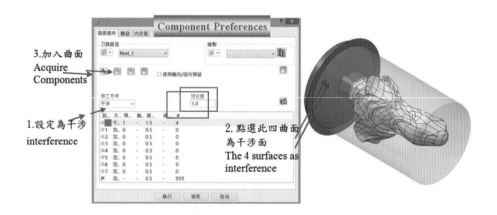

- 執行模型餘料加工的工法，計算後路徑如下，Calculate to obtain the tool path:

右側、前側、後側餘料加工：Rest_2、Rest_3、Rest_4 路徑，請參考上述作法或複製路徑 The stock model machining for the right, front and back are in the same manner or copy toolpath.

- 需要切換選擇不同的作動工作座標，Need to change Work coordinates-(L/F/B);
- 需要切換（L/F/B）重新運算安全高度，Recalculate Rapid Move Heights under work coordinates-(L/F/B);

計算後路徑如下，Calculate to obtain the tool path:

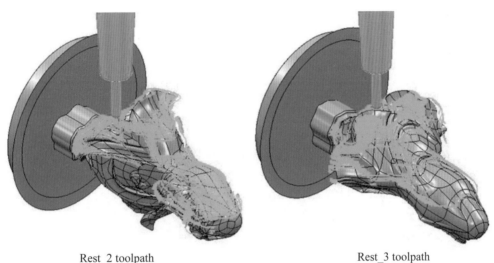

Rest_2 toolpath

Rest_3 toolpath

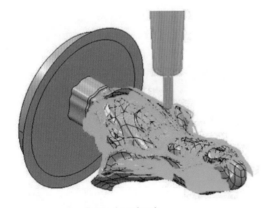

Rest_4 toolpath

* 運算完每條餘料加工的路徑，請加入殘料模型中做殘料運算，以產生後續最有效率的加工路徑，避免造成空切。After each "Model Rest Clearance" machining, the tool path has to be added into the stock model and calculate, to produce the most efficient tool path and ensure there is no "waste-run".

三、中銑 Semi-Finish Machining

- 模型右鍵功能中點選輸入參考曲面，Right click "Model" and import reference surface.

- 經由光碟 Chapter-08 輸入參考面：載入後並選取參考面

  From CD Chapter-08, Select and import reference surface: **"refer surface-finishing.dgk"**

- 工法選單選擇 沿面投影加工，Select **"Surface Projection Finishing"**;
- 路徑名稱，Toolpath Name: **Semi**
- 無需作動任何工作座標－(None)，No need to select any Work Coordinate;
- 選擇球刀，Select Ball Nosed tool **BN6**;
- 限制模式最小值，Setup Limit, Minimum: 7,
- 定義提刀高度（切換 None 重新運算安全高），Re-Calculate Rapid Move Heights;
- 定義進退刀與連結，進退刀可選擇曲面法向圓弧／角度／半徑，短連結可選擇圓弧連結，Define leads and links: "Surface Normal Arc/Angle/Radius" for Lead in/out; "Circular Arc" for Short link.

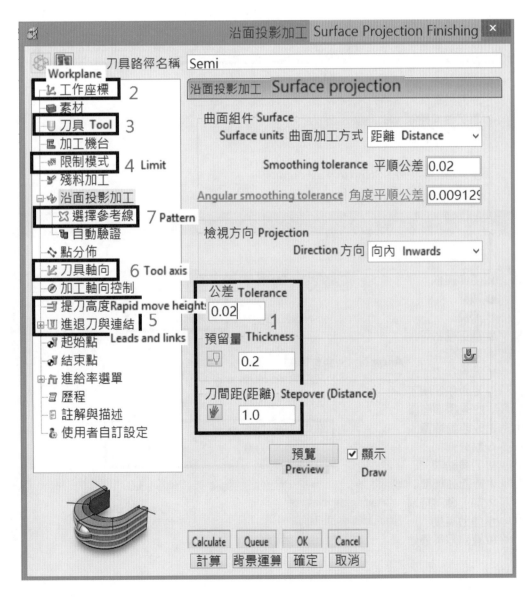

- 定義刀具軸向，Setup Tool-Axis

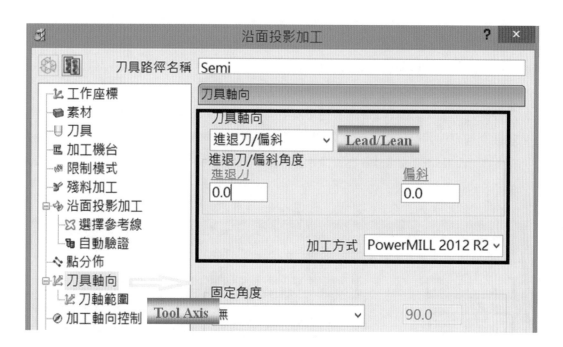

- 定義刀具軸向限制，Setup tool axis limits

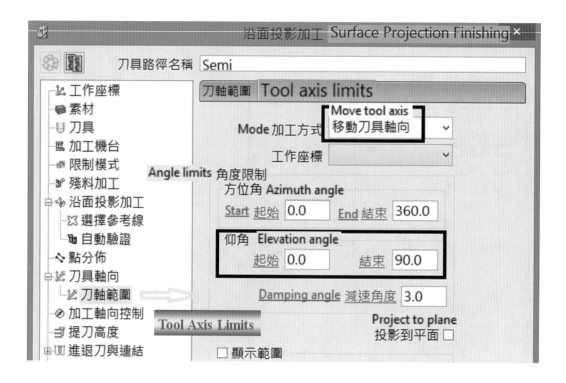

- 定義參考線投影的 UV 方向選擇，選擇螺旋狀功能，Define pattern direction,
  點選參考曲面之後，可點選預覽確認 UV 投影方向，click "reference surface" and
  visualize and confirm the direction of projection.

沿面投影加工 Surface Projection Finishing

刀具路徑名稱 Semi

選擇參考線 Pattern

- ⚒ 工作座標
- 🗋 素材
- 🗋 刀具
- 🖳 加工機台
- 🔧 限制模式
- 🗡 殘料加工
- 沿面投影加工
  - ✕ 選擇參考線 ⇨
  - 🗄 自動驗證
- ⟋ 點分佈
- 🗡 刀具軸向
- ◉ 加工軸向控制
- ⟍ 提刀高度
- 🗄 進退刀與連結
- 🗄 起始點
- 🗄 結束點
- 🗄 進給率選單
- 🗋 歷程
- 🗋 註解與描述
- 🗋 使用者自訂設定

選擇參考線
Pattern

Pattern direction
參考線方向 | V |

Spiral 螺旋狀 ☑

Ordering 路徑順序 | 單向 |

Start corner 起始角落 | 最大U最小V |

Sequence 補正順序 | 無 |

限制 (距離) Limits (Distance)

	☐ U	☐ V
Start 起始	0.0	0.0
End 結束	1.0	1.0

Preview
預覽

Draw
☐ 顯示

- 執行沿面投影中銑加工之前，需要在工具（Tools）> 顯示命令（Echo commands）輸入以下命令，Before "Surface Projection Finishing", the following commands have to be keyed in, under (Tools) > 顯示命令 (Echo commands):

EDIT SURFPROJ AUTORANGE OFF
EDIT SURFPROJ RANGEMIN -10
EDIT SURFPROJ RANGEMAX 10

　　此命令主要目的可將沿面投影距離限制在 +/- 10mm 範圍內，The purpose of the commands is to limit the projection distance within +/- 10mm。

　　（使用此命令那麼所選的參考曲面將無限距離做投影，導致刀具干涉無法產生路徑，Without these commands, the projection distance would be too high to cause tool collision）

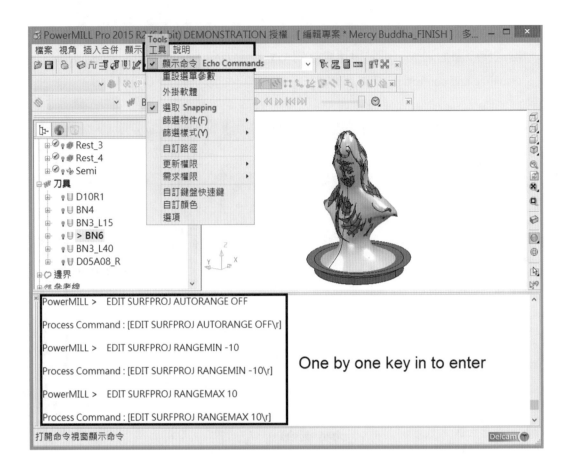

計算後路徑如下，Calculate to obtain the tool path:

8.3　精加工（Finish Machining）

一、底座五軸清角（Corner finishing Machining for the Base）

• 輸入參考面：**經由光碟 Chapter-08** 載入後並選取參考面，From CD Chapter-08, select and import reference surface：**refer surface-r1.5.dgk**

　　因為此底部區域以正面分析是倒勾整圈區域，五軸路徑同樣與三軸有其投影的方向限制，當路徑的產生無法依據軸向來投影產生理想路徑時，通常需借助工法策略與 CAD 建立參考面來做補助投影。Because the bottom range has overhang, like 3-Axis milling, 5-Axis still has limitation on projection. Therefore, special tool path strategies combined with CAD produced reference surfaces are to be used for assistant projection.

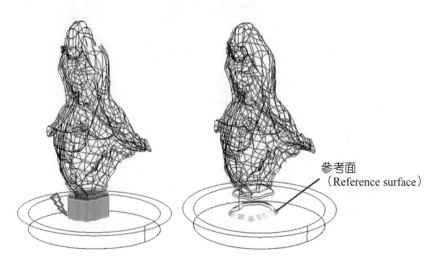

參考面
（Reference surface）

• 工法選單選擇 沿面投影加工或複製中銑 Semi 路徑，Select "Surface Projection
Finishing" Or copy "**Semi**" **toolpath**

複製 Semi 路徑只需更改以下選項，To copy tool path "Semi":

• 路徑名稱，Tool path **Name: Rest corner machining**

• 選擇球刀 Select Ball nosed tool: BN3;

• 定義刀具軸向限制（固定 30°），Setup tool axis limits (Fixed 30°)

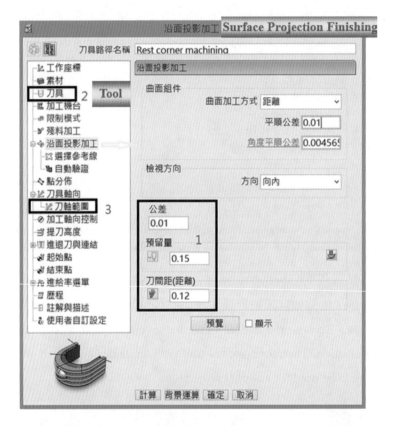

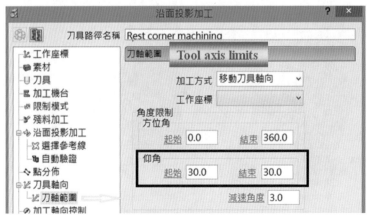

• 執行沿面投影五軸清角加工，計算後路徑如下，Calculate to obtain the tool path:

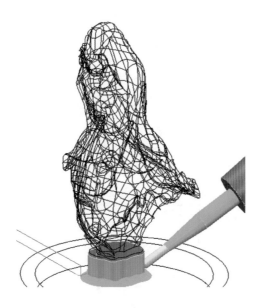

二、細銑 Finish Machining

• 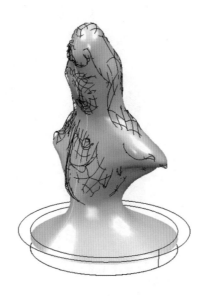 複製中銑 Semi 路徑，**Copy "Semi" toolpath**,

複製 Semi 路徑只需更改以下選項，Copy toolpath "Semi" procedure:

• 路徑名稱 Toolpath **Name: Finish**;

• 選擇球刀 Select ball nosed tool **BN3**;

• 選取同樣的參考曲面 Select reference surface - **finishing.dgk**

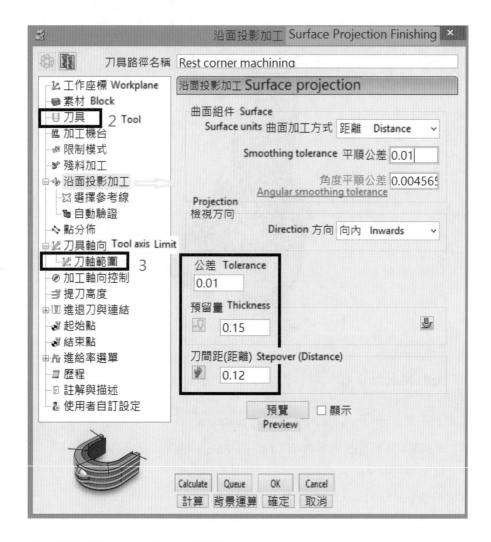

執行沿面投影細銑加工，計算後路徑如下，Calculate to obtain the tool path:

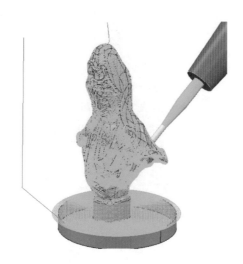

三、Logo 刻字加工（Scribing the Logo）

- 使用法線投影工法主要目的在於線段 Logo 文字可做五軸的法向角度加工，但需要將一般的參考線轉換成法線參考線（與曲面有 normal 關聯）才可選用，The "**Embedded Pattern Finishing**" is used to scribe the Logo letters, but need to transform regular pattern into normal reference curves.

操作如下，The procedure is as the foollowing:

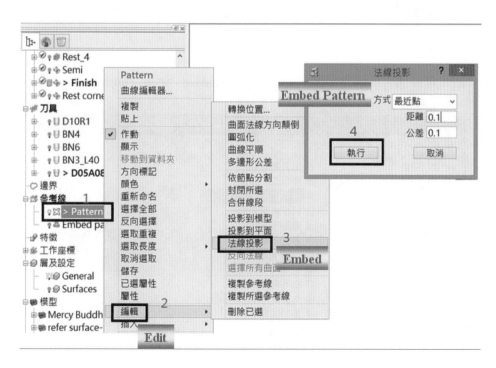

為避免加工的順序亂跳，可將參考線做順序優化編輯，Make patterns in good order to avoid tool paths out of order.

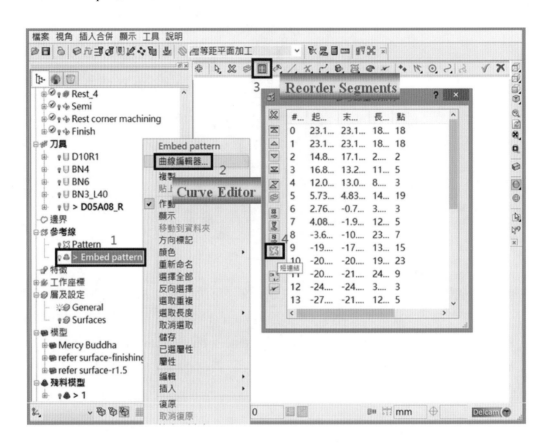

• 工法選單選擇  法線投影加工，Select "Embedded Pattern Finishing" strategy

- 路徑名稱 Tool path Name: Logo machining
- 選用錐形圓鼻刀，名稱 Select Ball Mill of 4mm diameter, Name: **D4_R**;
- 定義進退刀與連結，Define Lead and Links：進退刀設定無，Lead-in and Lead-out as（None），短連結選擇相對值 Short link as (Skim).

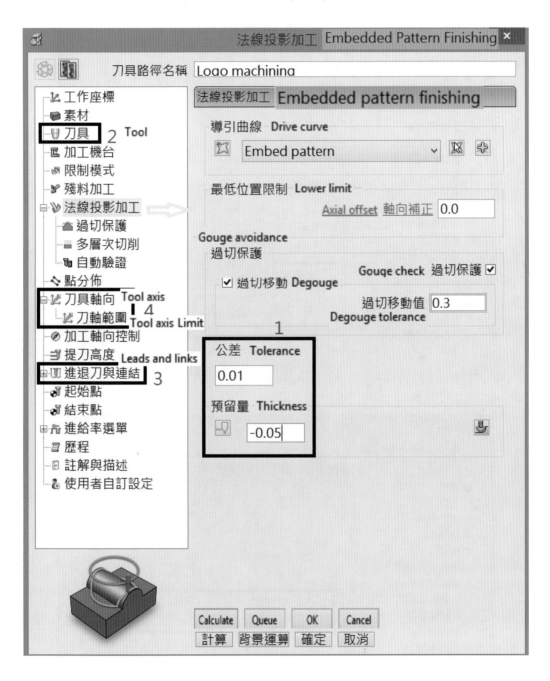

• 定義刀具軸向，Setup Tool-Axis

• 定義刀具軸向限制，Setup tool axis limits

執行法線投影加工，計算後路徑如下，Calculate to obtain the tool path:

　　所有路徑計算完後，務必執行干涉檢查，確保路徑安全性，Now overall tool paths are completed. Before export NC program for machine center, interruption checkups have to be conducted to prevent any collision in real machining.

　　下圖為完成後的觀音像。As the example, the completed statue of Guanyin Goddess of Mercy is show in the following pictures.

成後的觀音像，Completed statue of Guanyin, Goddess of Mercy

五軸加工範例：高腳杯
（5-Axis Macning: High-Foot Cup）

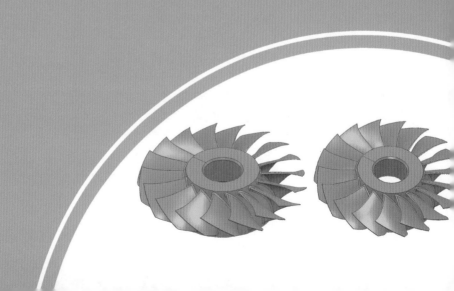

✎ 9.1 基本設定（Basic Setup）

一、輸入模型，Import Model: High-Foot Cup（高腳杯）

經由光碟 Chapter-09 輸入模型，From CD Chapter-09, import model: High-Foot Cup.

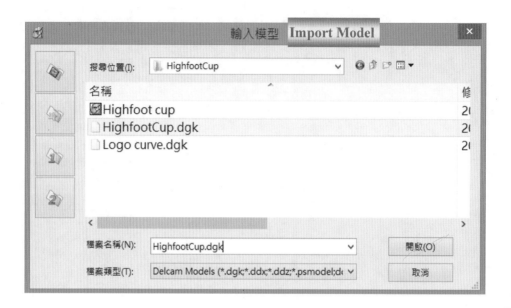

二、素材定義（Setup Block Material）

使用圓柱素材選項，Setup cylindrical material，直徑設定為 diameter：Ø61mm；高度最大值加大 1mm, Add 1mm to the height; In "Global Transform" or "The World coordinate."

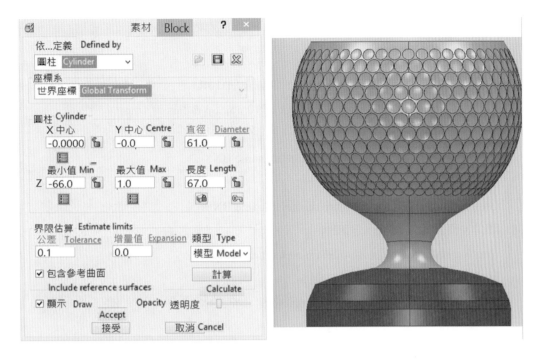

三、工作座標（Work Coordinate）

依照高腳杯的前、後分別建立不同 Z 方向的工作座標（Y 軸方向一致），Establish four work coordinates, with Y in the same orientation.

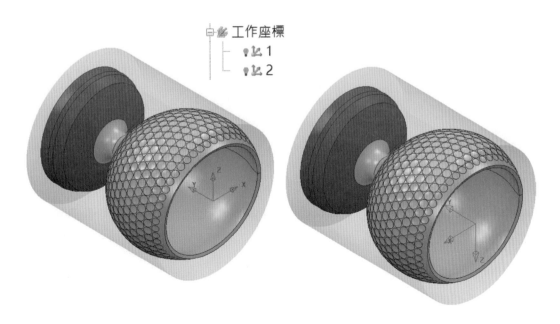

四、建立刀具（Create Tool）

- 刀具設定 Crete several tools:

1. 圓鼻刀，直徑 D10 鼻端圓角半徑 R1；名稱：**D10R1**,

   Create a Tip radiused tool with diameter of 10mm, tip radius of 1mm.

2. 球刀，直徑 8mm，名稱：BN8

   Create a Ball nosed tool of 8mm diameter, name BN8

3. 球刀，直徑 6mm，名稱：BN6

   Create a Ball nosed tool of 6mm diameter, name BN6

4. 球刀，直徑 4mm，名稱：BN4

   Create a Ball nosed tool of 4mm diameter, name BN4

5. 球刀，直徑 0.5mm，名稱：BN0.5

   Create a Ball nosed tool of 0.5 mm diameter, name BN0.5

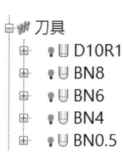

$\mathscr{J}_\odot$ 9.2   內側加工（Machining the Inside）

一、粗銑（Rough Machining ）

工法選單選擇 🖉 模型粗加工，路徑名稱 In Rough;

Create tool path: "Model Area Clearance" machining for left-hand side, Name: **In Rough**

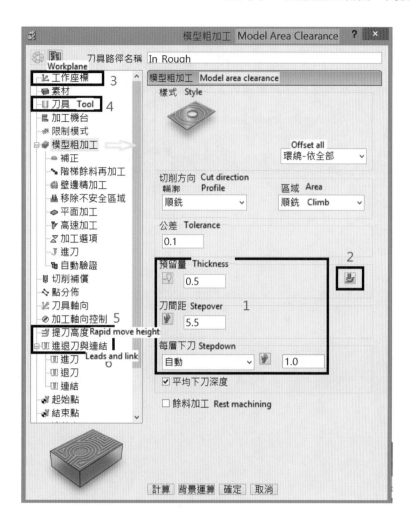

- 經由不等預留的選項功能定義干涉的曲面，Select "**Default thickness**" ．
  定義干涉曲面的用意在於避免加工到外部區域，The definition of the single surfaces as collision is to avoid machine outside region.

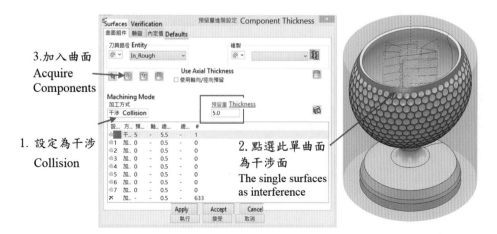

- 選擇世界工作座標，Select World Work coordinate;
- 選擇圓鼻刀 Select Tip radius tool: **D10R1**;
- 定義提刀高度（直接重新運算安全高），Define Rapid Move Heights ;
- 定義進退刀連結，Setup for leads and links ;

- 定義進刀爲斜向下刀，Setup "Lead" in for "Ramp"

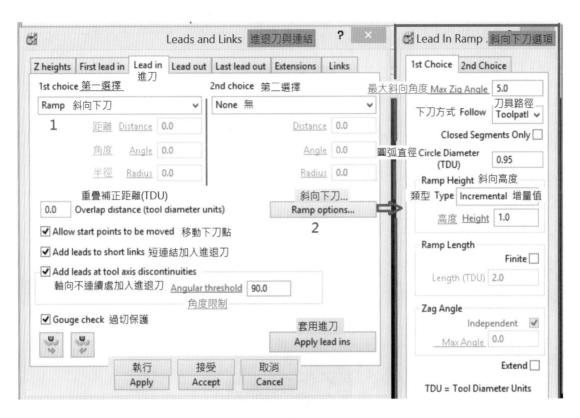

• 定義開始和終止點 🛢️，Setup for "Start and End point" 🛢️

• 執行模型粗加工的工法，前側計算後路徑如下，Calculate to obtain the tool path:

二、中銑（Semi-Finish）

• 工法選單選擇 線投影加工 Toolpath Stratage: "Line Projection Finishing"
• 路徑名稱 Toolpath **Name**: Semi
• 選擇球刀 Select Ball Nosed tool BN8

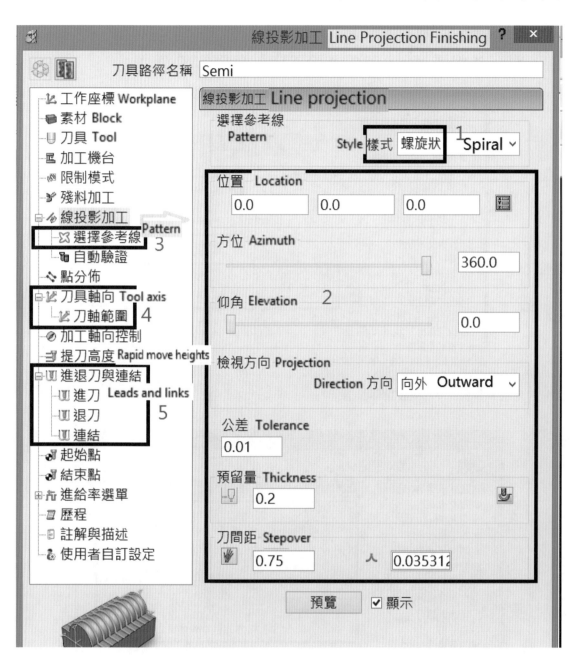

- 選擇參考線投影的限制範圍，選擇螺旋狀功能，點選預覽確認投影的位置與範圍是否正確，Select the limits of pattern projection and preview to confirm.

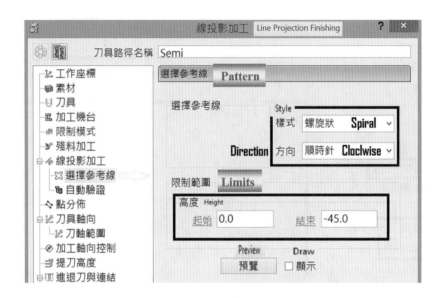

- 定義刀具軸向，Setup Tool-Axis

• 定義刀具軸向限制，Setup tool axis limits

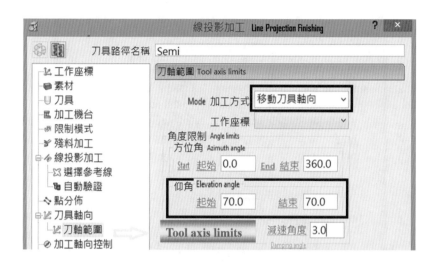

　　由於是加工內側曲面，因此線投影的檢視方向為向外。而刀具軸向則從中心線 Z 軸為基準朝外偏斜 20°（仰角 70°）。Because the inside surface is machined, the viewing direction of the line projection is toward outside, and the tool-axis rotate 20° outside (Elevation angle 70° degrees).

• 定義進退刀與連結，進退刀可選擇曲面法線圓弧／角度／半徑，短連結可選擇圓弧連結，Setup "Lead in/out" for "Surface normal arc".

• 定義連結為圓弧，Setup "Links" for "Circular arc"

• 執行線投影中加工的工法，計算後路徑如下，Calculate to obtain the tool path:

①由於粗銑路徑為三軸路徑，因此內側壁之處，還有留料約 4mm，因此中銑路徑 Semi 的預留量為 0.2mm，必須再複製運算補充兩個路徑，分別為 Semi_1 預留量 1.5mm, Semi_1_1 預留量 3mm。

Because the rough cut is 3-Axis machining, there is still 4mm material reminds inside. Therefore after the toolpath **Semi** (0.2mm thickness), two toolpaths have to be copied with thickness 1.5mm for toolpath: "**Semi_1**", and thickness 3mm for toolpath: "**Semi_1_1**".

②將這三條路徑合併方法如下，The procedure to append the three toolpaths:

將預計先行加工的 Semi_1_1 複製一份並更名為 Rest semi, 接著在 Semi_1 上，用滑鼠左鍵 +Ctrl 拖動到 Rest semi 時出現 + 號，放開左鍵出現詢問是否要合併路徑的對話框，回應（是）後，則同樣的方法將 Semi 合併到 Rest semi。

First, copy "**Semi_1_1**" into "**Rest semi**". Then with +Ctrl key pressed, drag "**Semi1**" to "**Rest semi**" until "+" sign occur. After release the left mouth button, answer "Yes" to the "Append" dialog box. Same manner to append **Semi** into **Rest semi**.

③合併之後的路徑 Rest semi 的連結需要重新計算檢查連結之後符號會顯示過切，此為預留量的因素實屬正常，加工無問題。

The combined toolpaths need to be recalculated and check. Although an "over-cut" sign appears, there is no real problem because it is caused during the "Thickness" calculation.

④計算後的路徑如下，Calculate to obtain the following Toolpath.

三、細銑（Finish）

- 經由光碟 Chapter-09 輸入參考面：載入後並選取參考面

  From CD Chapter-09, select and import reference surface: **Refer surface1.dgk**

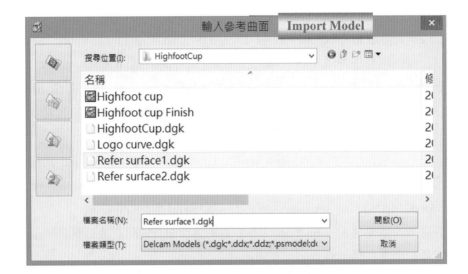

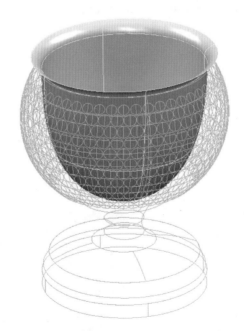

- 工法選單選擇 沿面投影加工 Toolpath Stratage: Surface Projection Finishing
- 路徑名稱 Toolpath **Name**: Finish
- 選擇同樣球刀 BN8; Select same Ball nosed tool BN8

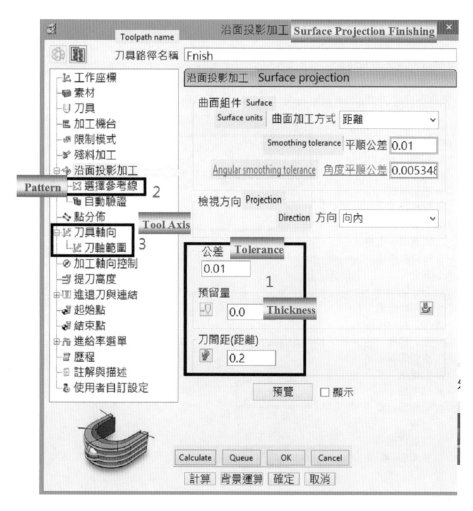

• 定義參考線投影的 UV 方向選擇，選擇螺旋狀功能，點選參考曲面之後，可點選預覽確認 UV 投影方向。Define pattern direction V and Spiral. After select the reference surface, you can click "Preview" to confirm the direction of projection.

- 定義刀具軸向，Define Tool-axis

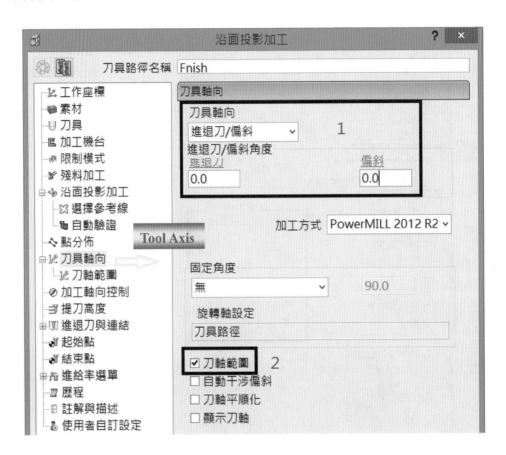

- 定義刀具軸向限制，Setup tool axis limits

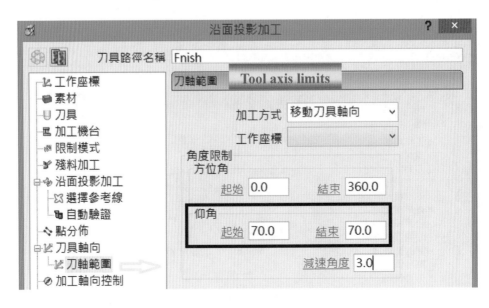

- 執行沿面投影中銑加工之前，需要在工具 > **顯示命令**，輸入以下命令，Before the "Surface projection finishing", in Tools> Echo commands: Key-in the following commands:

  EDIT SURFPROJ AUTORANGE OFF

  EDIT SURFPROJ RANGEMIN -10

  EDIT SURFPROJ RANGEMAX 10

此命令主要目的可將沿面投影距離限制在 +/- 10mm 範圍內。（無使用此命令那麼所選的參考曲面將無限距離做投影，導致刀具干涉無法產生路徑）。This is to limit the "Surface Projection" distance within +/-10mm. Without this limitation, infinite distance will occur in projection, so that tool interference will happen and the software will fail to calculate toolpath.

- 執行內側沿面投影加工的工法，計算後路徑如下，Calculate to obtain the tool path:

9.3 外側加工（Machining the Outside）

一、粗銑（Rough Machining）

- 工法選單選擇 模型粗加工，路徑名稱 **HF_Rough**；

Create tool path: Rough machining "Model Area Clearance" for left-hand side, Name: **HF_Rough**

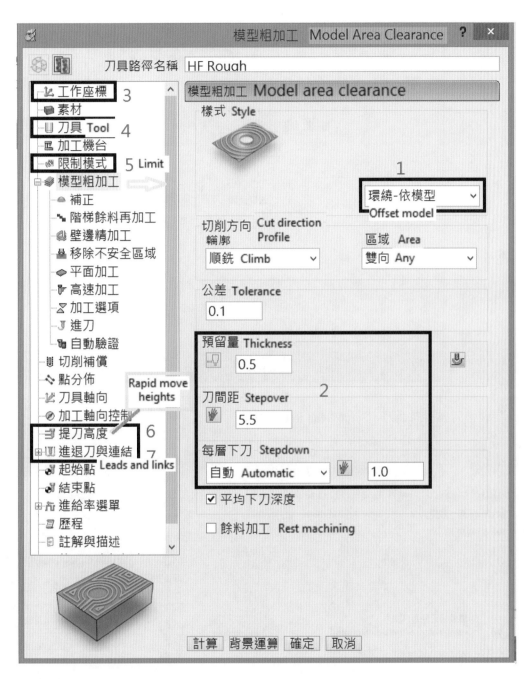

- 選擇作動工作座標，Select Work coordinate - (1);
- 選擇圓鼻刀，Select Tip radius tool, Name: D10R1
- 限制模式，最小值 Select Limit, minimum: -2 ,
- 定義提刀高度（切換1座標重新運算安全高），Calculate Rapid Move Heights under WC 1;
- 定義進退刀連結，Setup for leads and links 🔲 ;

• 定義進刀為斜向下刀，Setup Lead in for Ramp

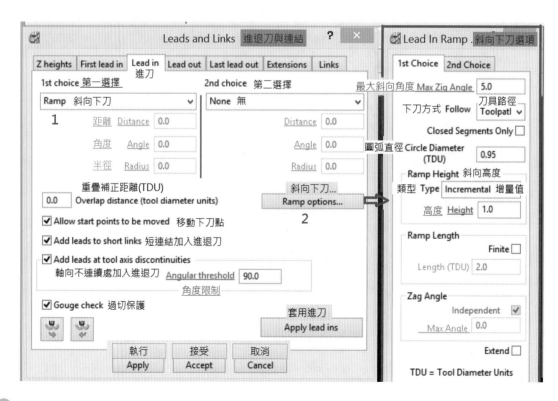

• 定義開始和終止點，Setup for Start and End point

• 執行模型粗加工的工法，前側計算後路徑如下，Calculate to obtain the tool path:

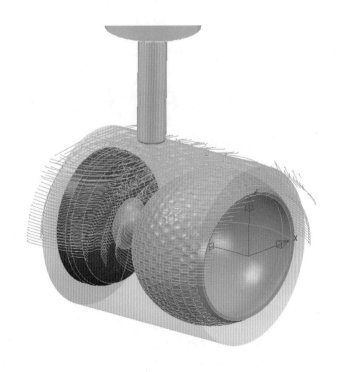

後側粗加工：HB_Rough 路徑，請參考上述作法或複製 HF_Rough 路徑，Machining for the back side is in similar manner or copy HF_Rough toolpath.

- 需要切換選擇不同的作動工作座標，Need to change Work coordinate - (2);
- 需要切換 (2) 重新運算安全高度，Calulate Rapid Move Heights under WC2;
- 計算後路徑如下，Calculate to obtain the tool path:

二、中銑殘料加工 Stock for semi

- 經由光碟 Chapter-09 輸入參考面：載入後並選取參考面，From CD Chapter-09, Select and import reference surface: **Refer surface2.dgk**

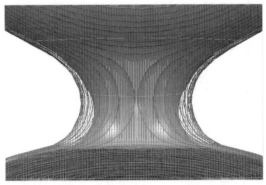

頸部殘留料，Remaining stock model at the neck

三、中銑（Semi-Finish Machining）

- 工法選單選擇 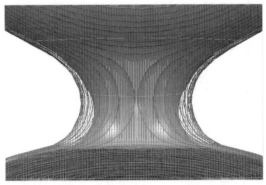 沿面投影加工，Toopath Stratage: **Surface Projection Finishing**
- 路徑名稱，Toolpath name: **Stock semi_1**
- 無需作動任何工作座標 - (None)，Don't select any Work coordinate;
- 選擇球刀，Select Ball nosed tool: BN6
- 限制模式，Setup Limit: 最大值 Maximum -44，最小值 Minimum -50.5,
- 定義提刀高度（切換 None 重新運算安全高），Calculate Rapid Move Heights;
- 定義進退刀與連結，進退刀可選擇垂直／角度／半徑，路徑切線延伸 15mm,
  短連結可選擇圓弧連結，Define Lead and Links: "Vertical/Angle/Radius" for Lead in/out;
  Extension 15mm; "Circular arc" for Short link.

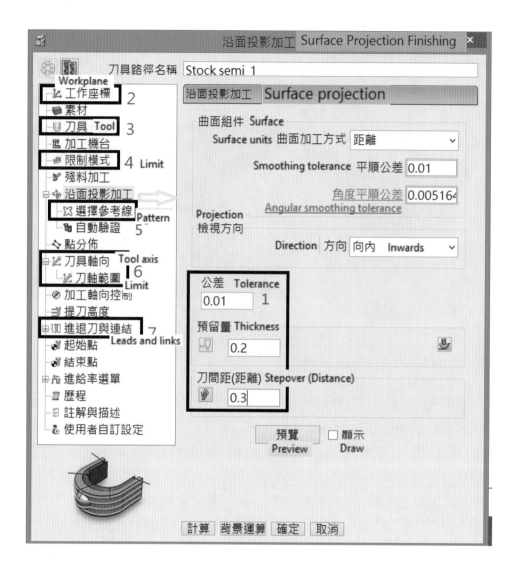

- 定義參考線投影的 UV 方向選擇，選擇螺旋狀功能，限制 V 投影範圍，Define pattern direction V as spiral. Check on V for limits.
- 點選參考曲面之後，可點選預覽確認 UV 投影方向。After select the reference surface, you can click "Preview" to confirm the direction of projection.

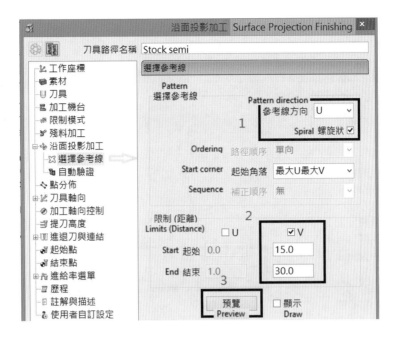

• 定義刀具軸向，Setup Tool-Axis

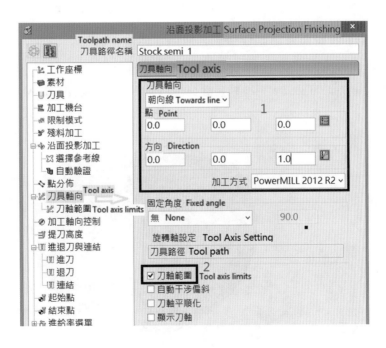

五軸銑削數控加工之基礎及實作

- 定義刀具軸向限制，Setup tool axis limits

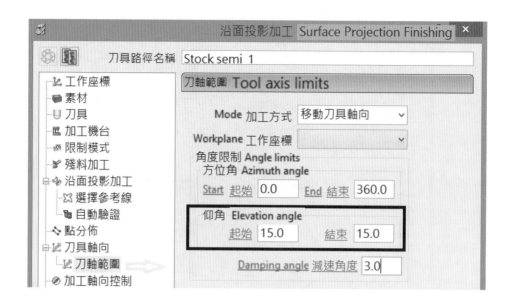

運算路徑前請從命令列，Before calculate the toolpath, please key in:

EDIT SURFPROJ AUTORANGE ON

此命令可將**沿面投影**距離範圍恢復到系統內定設定（無限制）。This command is to recover the "Surface projection" distance to the default value (no limit).

- 執行沿面投影加工的工法，中銑殘料計算後路徑如下，Calculate to obtain the tool path:

- 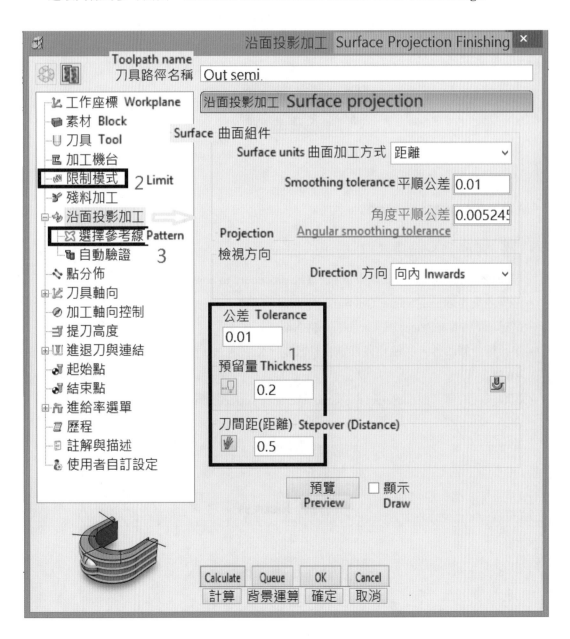 複製中銑殘料加工路徑 Copy toolpath "Stock semi_1";

  複製 Stock semi_1 路徑只需更改以下選項，The procedure to copy "Stock semi_1":

- 路徑名稱，Toolpath name: **Out Semi**

- 同樣選擇球刀，Select Ball nosed tool: BN6

- 取消限制模式 - 最大值 -44，最小值 -50.5 Cancel Limit;

- 取消不勾選限制 V 投影範圍，Un-check limit V in pattern set up.

- 選取同樣的參考曲面，Select the same reference surface: **Refer surface2.dgk**

• 執行沿面投影加工的工法，中銑計算後路徑如下，Calculate to obtain the tool path:

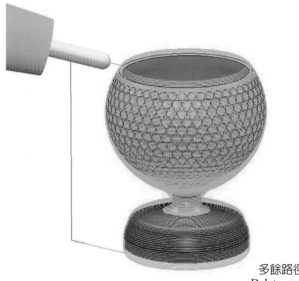

多餘路徑請編輯刪除
Delete excessive toolpath

四、細銑（Finish Machining）

• 複製中銑 Semi 路徑，Copy the semi-finish toolpath;

  複製 Semi 路徑只需更改以下選項，The procedure to copy toolpath Semi:

• 路徑名稱 Toolpath Name: Out finish;

• 選擇球刀 BN4; Select Ball nosed tool **BN4**;

• 除預留量修改外，其餘相同參數條件設定，Keep same parameters, except for **Thickness = 0.0**;

• 選取同樣的參考曲面 Select the same reference surface: **finishing.dgk**

- 執行沿面投影加工的工法，細銑計算後路徑如下，Calculate to obtain the tool path for the finish machining:

多餘路徑請編輯刪除
Delete excessive toolpath

五、刻字 Logo 加工（Machining of the Logo）

你可參考足球章節來自行產生加工路徑，To machine the loge, you can refer chapter-7 "Soccer Ball".

所有路徑計算完後，務必執行干涉檢查，確保路徑安全性

After completed all the tool paths and before export NC program for machine center, interruption checkups have to be conducted to prevent any collision in real machining.

五軸加工範例：渦輪葉片
（5-Axis Machining: Turbine Blades）

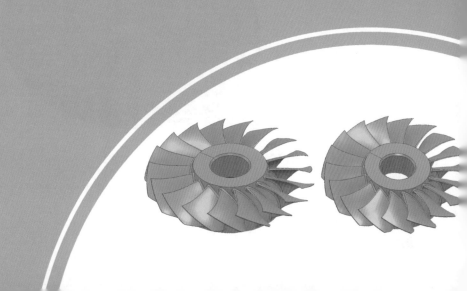

10.1 葉片工法及前置作業 （Turbine Blade Machining Strategies）

在工法選單中，對渦輪葉片有三種專用工法。In the "Strategy Salector" dialog box, add three tool path strategies specifically designed for blades.

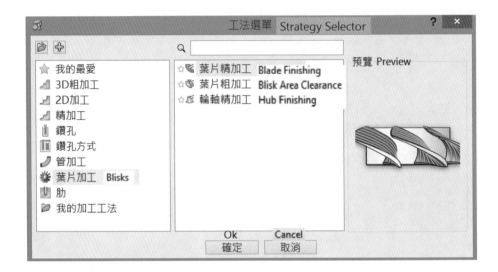

- 選擇檔案 – 刪除所有，Delete all
- 選擇工具 – 重設選單參數，Re-set parameters
- 經由光碟 Chapter-10 打開專案，Open project from CD Chpater-10: **BliskSimple_Start**;

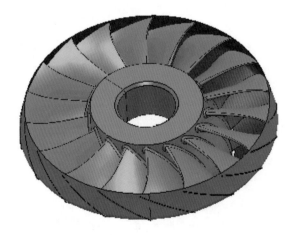

- 作動工作座標 **1**，Activate Work Coordinate #1.
- 根據模型大小產生**圓柱**素材，Create cylindrical block material 。

- **重設快速提刀高度**，Set "Rapid Move Height" ⬛。
- 設定**開始點 / 結束點**爲第一點 / 最後一點安全高 **Z**, Set "Start/End" 🔲。
- 在**進退刀連結**選單設定爲**相對提刀高度**，Setup "Lead & Link" 🔲。
  爲了能夠順利執行渦輪葉片加工方案的組成，部分表面必須先取得一系列特殊命名的層別。在這個例子中，輸入模型的曲面已經獲得適當指名的層別。To smoothly plan the machining procedure for the turbine blades, some surfaces should be put into specially named layers. In this example, the surfaces were already in the layers for this imported model.

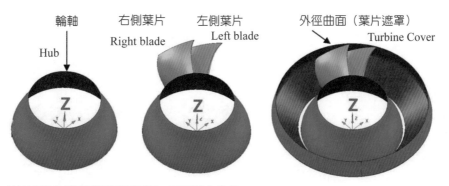

輪軸 / Hub　右側葉片 / Right blade　左側葉片 / Left blade　外徑曲面（葉片遮罩）/ Turbine Cover

註 Note：葉片遮罩的外型縱斷面需平順，不可以有尖角，The surface of the turbine cover must be smooth and no sharp corners are allowed.

- 作動刀具爲球刀，Activate the ball nose toll BN25.
- 打開預留量進階選單 🔲 並選擇內定值，Open "Component Thickness" and select Default value.
- 選單內點選某一個顏色組別，在點選模型的**外徑曲面**並按「**取得組件**」按鈕 🔲，
  Select the cover surface and click "Obtain components" 🔲, the first row appears.
- 點選「**忽略**」並按「**執行**」，之後按「**接受**」並跳出選單，Apply and Accept.

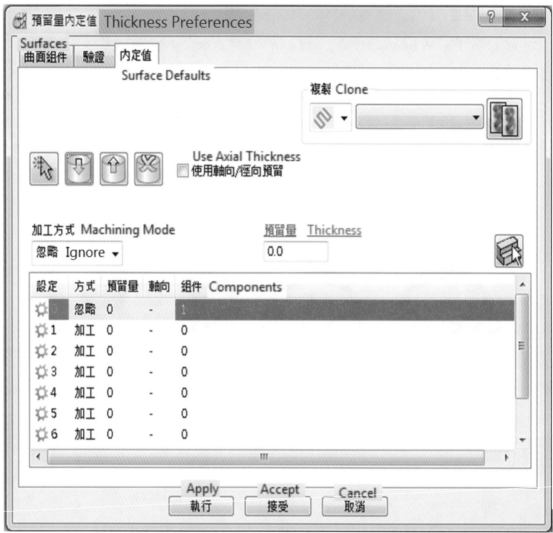

10.2 葉片粗加工（Blade Rough Machining）

在路徑工法選單中選擇葉片加工，點選葉片粗加工選單，Select "Blisks Area Clearance" toolpath strategy.

• 輸入正確數值如下，在**圓角處曲**面及分隔葉片選擇空白即可，Input values shown in the following picture.

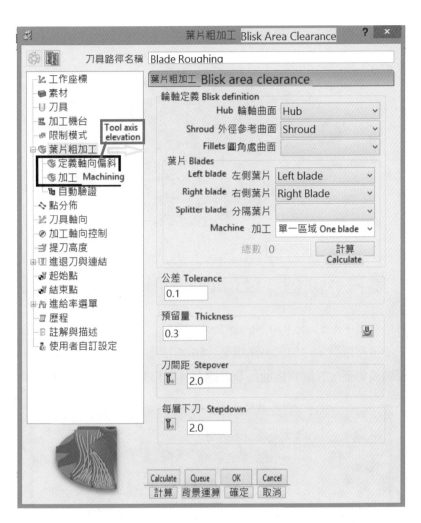

• 定義軸向與加工方式，Define Tool-Axis elevation and Machining parameters

- 執行葉片粗加工的工法，計算後路徑如下，Calculate to obtain the tool path:

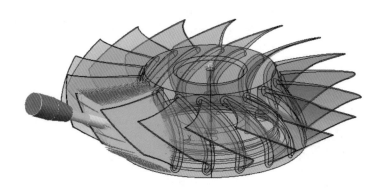

10.3　葉片精加工（Blade Finishing）

- 作動刀具爲球刀，Activate Rall Nosed Tool **BN15**.
- 在路徑工法選單選擇葉片加工，點選**葉片精加工**選單，Select "Blade Finishing" toolpath.
- 輸入正確數值如下，在**圓角處曲面**及**分隔葉片**選擇空白即可，Input values as shown:

• 定義軸向與加工方式，Define Tool-Axis elevation and machining direction

定義軸向偏斜 **Tool Axis elevation**

定義軸向偏斜

從 | 輪軸曲面法向 | **Hub Normal**

角度 0.0

加工 **Machining**

加工

Machining Direction 切削方向 | 順銑 | **Climb** | ∨

• 執行葉片精加工的工法，計算後路徑如下，Calculate to obtain the tool path:

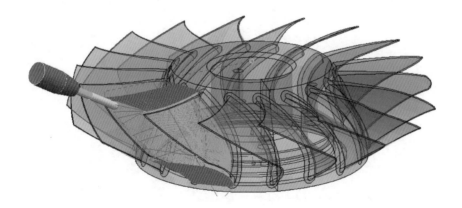

10.4　輪軸精加工（Hub Finishing）

• 在路徑工法選單 選擇葉片加工，點選**輪軸精加工**選單，Select "Hub Finishing" tool path
• 輸入正確數值如下，在**圓角處曲面**及**分隔葉片**選擇空白即可，Input values as following

五軸銑削數控加工之基礎及實作

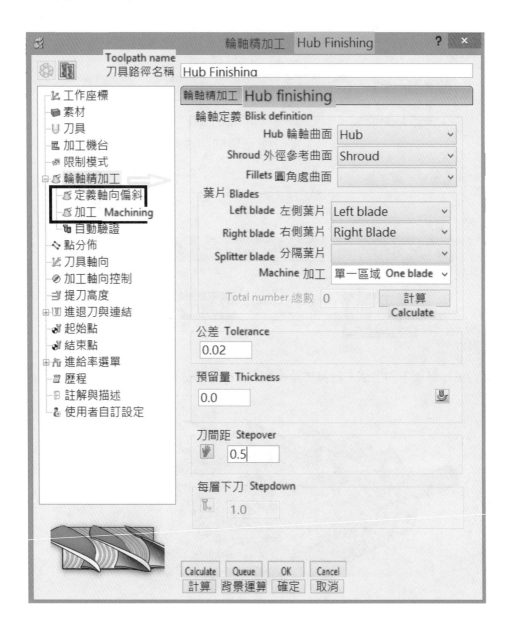

• 定義軸向與加工方式，Define Tool-Axis elevation and Machining direction

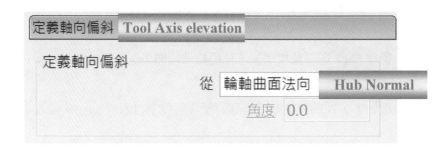

• 執行輪軸精加工的工法，計算後路徑如下，Calculate to obtain the tool path:

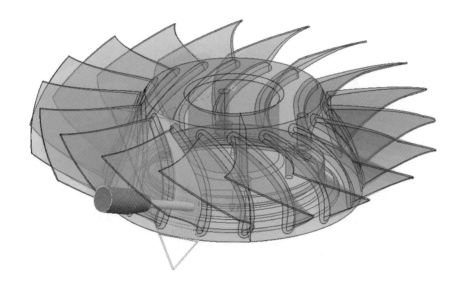

10.5　加工模擬擬（Simulation）

• 葉片路徑都產生後，執行 **ViewMILL 模擬器**，After toolpath calculated, you can simulate the tool path with **ViewMILL**.
執行 ViewMILL 之前，需要定義素材並先行產生三角形模型，外徑曲面是銑削前的材料。Before ViewMILL, a block material must be defined by the outside surface.

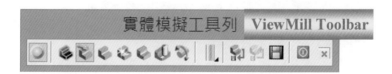

• 在素材選單選擇**加工模型**（形狀素材），並點擊由**檔案載入素材**，To define the block material, use "Machining Model" and input the file for the block.

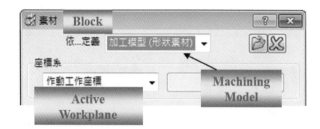

• 經由光碟 Chapter-10 選擇三角網格模型檔案，Select triangle model:-**TurnedShroud-EX1.dmt**

• 開始 **ViewMILL** 模擬器後，點選實體模擬開／關 ICON，Click the "ViewMill on/Suspend" button 🔵 to turn on Simulation.

- 路徑 Blade Roughing 用滑鼠右鍵點擊**刀具起始位置**，並在模擬工具列按**開始模擬**，
  Right-click toolpath "Blade Roughing", select "Starting Position" and play simulation.

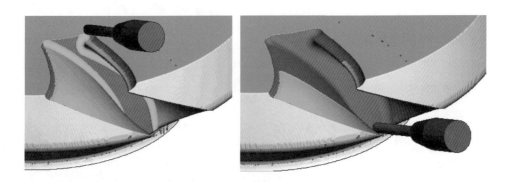

- 路徑 Blade Finishing 用滑鼠右鍵點擊**刀具起始位置**，並在模擬工具列按**開始模擬**，
  Right-click toolpath "Blade Finishing", select "Starting Position" and play simulation.

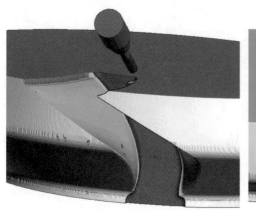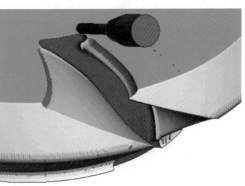

• 路徑 Hub Finishing 用滑鼠右鍵**點擊刀具起始位置**，並在模擬工具列按**開始模擬**，
Right-click toolpath "Hub Finishing", select "Starting Position" and play simulation.
你可切換材質為金屬做模擬了解 ，You can also switch to "Metal" to view.

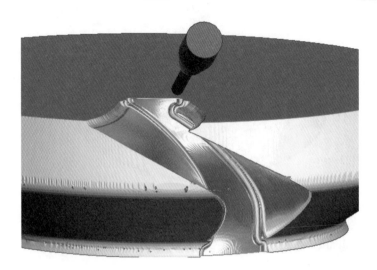

• 繞 Z 軸每 20 度來複製路徑這些葉片的路徑，並繼續**模擬**，You can copy the toolpaths
about Z-Axis, in 20 Degree interval to complete all the blade machining.
完成後 ViewMILL 的影像如下，The finished machining is as the following.

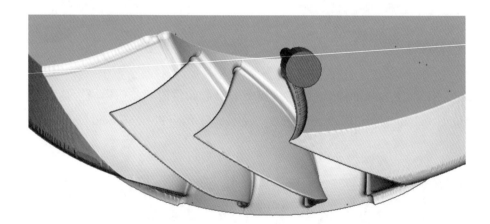

加工機台模擬
（Machine Simulation）

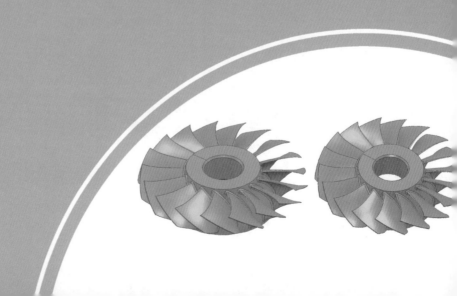

PowerMILL 軟體提供**機台 - 模型**之間的**干涉驗證**，透過機台模擬工具列，可針對多軸加工進行模擬，並提供使用者有效的干涉訊息。With "Machine Simulation" provided by CAM software, the collision problems can be detected and prevented, before the NC program is exported to machine centres.

使用**機台模擬模組**進行動態模擬時，於碰撞發生的位置，模擬將自動停止，並且機台干涉組件會呈現**紅色**在螢幕上出現一**警告訊息**，直到使用者確認後，系統會將發生碰撞的刀具路徑位置列表至加工訊息 - 機台干涉頁面中。During Machine Simulation, it will stop at collision and show a warning. It will continue after confirmed by the user and will list all the collision positions in "Machine Collision" table. 標準**機台**模擬純屬展示觀看，無干涉功能；若使用者需要仔細驗證是否干涉，可付費購買**機台模擬模組**，Standard version is for demonstration only. For real detection, commercial module can be purchased.

PowerMILL 安裝資料中提供了三個基本的多軸**機台模擬**（.mtd）檔，通常它位於硬碟 C 的以下目錄。或經由光碟 Chapter-11 的 "Machine Data" 資料夾內取得，There are three basic Multi-Axis simulation files:

C:\Program Files\Delcam\PowerMILLxxx\file\examples\MachineData

Or you can import the files from CD Chapter-11, attached with the book.

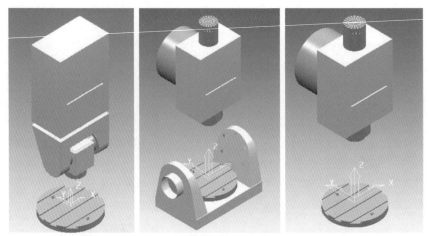

注意：機台模擬（.mtd）檔中的模型以及各個軸向移動範圍和界限都是透過實際使用機台時精確複製所得。由於設定的標準不同以及公差等等原因，每個機台模擬 (.mtd) 檔及其相關的模型都必須針對各個機台進行測試和精確調整。

Note: The models as well as moving/rotation limitations in machine simulation file (.met) are carefully measured for actual machines. The final version of the file must be tested and measured for each particular machine.

模擬步驟（Procedure of Simulation）

- 刪除全部，重設選單參數，Delete all and reset parameters,
- 經由光碟 Chapter-11 開啟專案 Open Project: Collision_Simulation\Swarf_Check,
- 使用滑鼠右鍵點擊樹狀列中的刀具路徑，Right-click the toolpath: Outer Swarf，並從功能表中選取刀具起始位置，Select "Tool Start Position".

於是路徑模擬工具列即出現在螢幕，The "Toolpath Simulation" will show as the following:

- 從下拉式功能表選取視角－工具列－機台模擬，開啟機台模擬工具列，Select View - Tool Bar – "Machine Simulation":

- 選取輸入機台模型按鈕並從目錄 Machine Data 選取 dmu50v.mtd。
  Click "Import Machine" Icon and select "dmu50V.mtd" from "Machine Data" directory.

通常將**機台模型**的原點（工作座標）定位在旋轉工作台中心的頂部，The Original machine origin is usually set at the top of rotary table. 作動 PowerMILL 刀具時，會自動對齊機台主軸，When the tool is activated, it will automatically align with the spindle of the machine center.

• 確認已點選顯示／不顯示觀台圖示 ![icon]，顯示機台
• 選取從前面查看（-X）並放大加工區域，
　Show Machine and view front.
• 選取模擬工具列中的以刀具視角觀看圖示。
• 選取模擬工具列中的加工訊息選項，
　View machine information.

加工資訊選單顯示刀具位置和干涉位置資訊，Machining information demonstrates tool position and collision.

選取**位置**頁面後即可顯示出機台位置。各軸向下方白框內的數值為**相對應位置**以及機台具有的 5 個軸向，**B 軸**和 **C 軸**是旋轉軸，而 **X 軸**、**Y 軸**和 **Z 軸**為線性軸，各軸向左右的數值顯示每個軸向的行程範圍，The active position is shown in "Machine Tool Position" window with all five-coordinate values, where X,Y,Z are linear and B, C are rotational, for this 5-Axis machine.

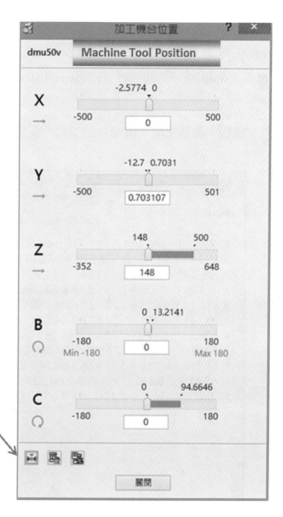

按下**數值歸零**按鈕，
可得到自指定原點的絕對值,
Absolut Zero

按下　可得知機台原點位置,
Machine Origin

按下　可得知碰撞位置頁面,
Opens Collision window

- 選取**機台干涉**頁面，Select "Machine Collision"
- 開始**模擬** ▷，直接觀察模型加工，Play and start simulation.

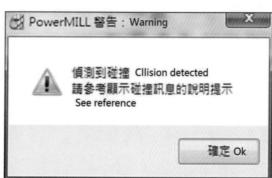

軟體發現干涉，螢幕上出現一**警告訊息**，Warning occurs if the software detects interruptions. 以上**警告**資訊僅在發生第一次碰撞時後出現，其他的干涉將在**機台碰撞**頁面顯示，Above warning only list at the first collision, other collisions will appear in the "Machine Collision" window.

- 點擊**確定**，繼續**模擬**。按顯示加工機台按鈕 ，Click ok and continue simulation.

- 從列表中選取**機台碰撞**的某一個位置，Select any machine collision in the list:

透過**模擬**將刀具直接移動至列表中所選的碰撞位置，就可以直接觀察該干涉情形，By moving to the position of collision in the list, you can visualize the collision situation first-hand.

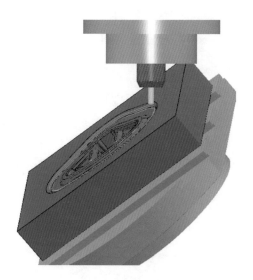

透過機台模擬可以清楚知道碰撞發生的位置和細節。使用者可立即採取對應措施來避免干涉。就本範例而言，我們只要把刀具長度做適量增長即可解決問題。Through machine simulation, the detail and the position of collision can be detected. Corresponding measures could be applied to avoid the collision. In the following example, the problem is solved by extension of the tool.

五軸銑削數控加工之基礎及實作

• 在**樹狀列**中使用滑鼠右鍵點擊刀具 **Tip Rad 10 3**，從選單中選取**設定參數**，Right-click the tool: **Tip Rad 10 3**, Select "**Set Parameters**" to change the tool parameter.

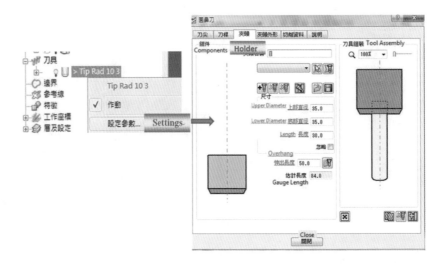

• 選取**刀具**選單中的夾頭頁面，將**伸出長度**修改為 50, Select Holder and modify to 50.

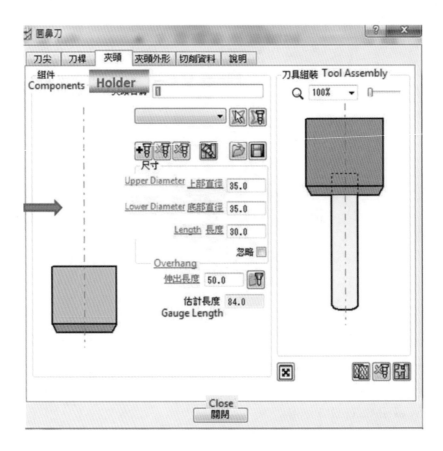

- 使用滑鼠右鍵點擊樹狀列中的**刀具路徑 Outer Swarf**，從功能表中選取**刀具起始位置**選項，Right-click toolpath **Outer Swarf**, Select "**Simulete from start**".

- 選取**模擬**工具列中的**打開加工訊息顯示** ，Select "Open Machining Info."
- 選取選單中的**機台碰撞**頁面，Select "Machine Collision" window.
- 點擊**清除**按鈕 | 清除 | 清除已有的碰撞顯示，Clean existing collision info.
- 開始**模擬** ▷，Start simulation.

加工機台碰撞 Machine Collision

dmu50v

檢查模式
| 清除 | ○ 無　　　　◉ 固定

Close | 關閉

此時**機台干涉**頁面仍然是空的，表示沒有發現干涉，Above window presents an empty box, that means "no-collision" has been detected.

加工機基本設定及運行
（Basic Setup and Start Machining）

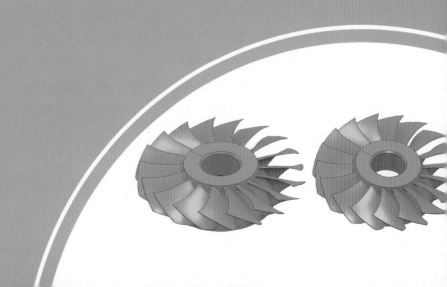

12.1　四軸銑床基本設定及運行實例-Roland MDX-40A CNC 雕刻機（Basic setup and operation for Roland MDX-40A 4-Axis Milling Machine）

　　本教材以 Roland MDX-40A 微小型 CNC 四軸加工機為例進行敘述。日本 Roland 公司依據市場對微小型 CNC 加工機種的反應與功能需求於 2009 年 7 月底推出更新的第二代機種 MDX-40A。此 MDX40A 充滿許多現代的新設施與功能，尤其是能搭配目前市面上任何先進的 CAM 系統，這些新的特色包含：機械式刀長感測器；支援標準 NC- 碼；以及改良過的旋轉第四軸 A 軸。由於可以支援 NC- 碼 /G/M 碼，MDX-40A 對於教育學校環境或是需要切削小型 3D 模型研發設計單位均是理想的加工平臺。改良後的旋轉軸現在已牢固地座落於工作平臺上，並且比原來 MDX-40 有更大的機械穩定度。此新旋轉軸也有較大與較長的夾持與工件空間，可允許加工較以往更大的物件。例如加大後的旋轉軸尺寸已可以完整加工一個 500cc 寶特瓶的模型。The Roland MDX-40A CNC machine combines the advanced capabilities of a Subtractive Rapid Prototyping® (SRP) system with the ease-of-use of a bench top CNC mill to provide product engineers and designers with a powerful 3D milling device. Offered at less than half the cost of additive systems, MDX-40A CNC mill produces affordable and highly accurate prototypes from a wide variety of non-proprietary materials with greater precision and excellent surface finish. Unlike other CNC machines used for rapid prototyping, the compact design of the MDX-40A is the perfect fit for any office and classroom space. It can even function as a CNC engraving machine.

　　此外 MDX-40A 還配備 PC-Based 操控面板 VPanel，允許使用者更迅速與容易地調整刀具位置與程式設定。使用此操控面板，用戶可以使用滑鼠操控面板上的垂直，水準與傾斜的方向鍵以移動刀具到任何需要的位置上。也可以調整以方向鍵移動機器的速度以便更精準方便的做工件原點設定。最後 MDX-40A 允許在機器加工同時時，仍可以手動調整功能（override function）調整到最適合加工的主軸轉速與切削速度。The MDX-40A supports an on-screen operation panel that allows you to adjust the location of the mill-tool and quickly program settings. Using this panel, you can move the cursor in vertical, horizontal and oblique directions and to the desired position for the most efficient tool path. You can also adjust the speed of cursor movements for easier origin setting.

　　以下為 Roland 4 軸加工機的基本設定及運行步驟。其他四軸加工機型及控制器在設定及運行上基本上大同小異。The basic setups is illustrated as the following. Similar steps will apply to other brand of 4-Axis CNC machines and controllers:

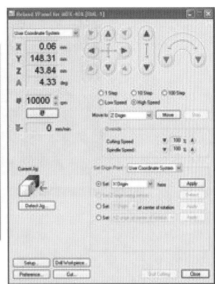

Roland 4 軸加工機以及虛擬控制台

The Roland MDX-40A 4 axis machine and Virtual-Panel

一、機器原點設定（Original machine setup for home position）

　　對於新購進機台或者 3 軸機新增第四軸需要進行原點設定。只需將標定中心杆安裝在 X 軸上，以及將測試頭裝在刀杆上，然後在虛擬控制板（V-Panel）上按下 "Detect jig" 即可自動設定。一般來講這個動作只需進行一次。We begin by calibrating the 4 axis unit. Assemble the calibration bar and the sensing tool on the machine. Make sure it is center bar and detection panel connected. Go to V-Panel and click "Detect jig" to calibrate home position of the 4 axis unit. This action should conduct only once.

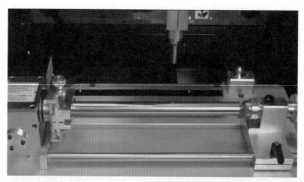

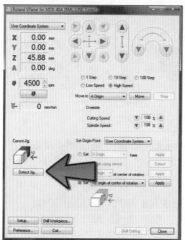 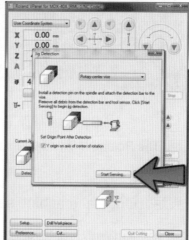

V-Panel 上操作 4 軸機原點設定（Calibrating the 4 axis unit on V-Panel）

二、加工零件前的座標及刀具長度設定（Setup for specific work piece）

若要加工一個新零件，則必須進行一下的設定，

To machine a 3D model, follow the setup steps:

第一步，Step 1：夾持工件並進行右端中心支撐，Clamp the material and make a center hole on the other side.

先利用左邊的夾爪將工件基本夾緊（下圖左），然後利用中心鑽在工件右端加工一個中心孔（下圖右以及 V-Panel 圖）。中心孔完成後，必須將中心鑽替換成中心支撐並進行固定。Mark the center of the block material on the right and clamp on the machine. Then drill a center hole with the center drill on the tail-stock. After drilling the hole remember to change the centering drill to the center support. The V-Panel operation for center hole drilling is shown in the pictures bellow.

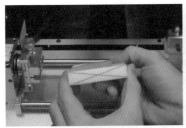 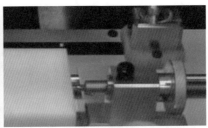

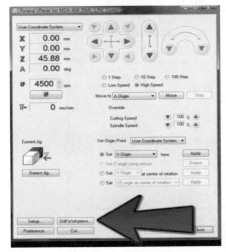 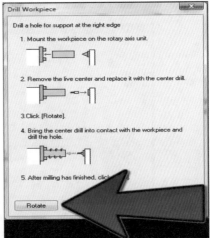

工件材料夾持及鑽中心孔（Clamp the block material and drill the center hole）

第二步 **Step 2**：刀長以及 X 原點設定，Setup for the length of the tool and x origin

刀長設定，**tool length setup:**

對於圓柱形材料：安裝刀具完畢後，在虛擬控制台對話方塊裡的點擊「Z 原點」自動測量設定刀長。For cylindrical material: assemble the tool and click "Z origin" in the V-Panel to set the tool length.

對於長方形材料：安裝刀具完畢後，在虛擬控制台對話方塊裡的點擊「YZ 原點」自動測量設定刀長。For rectangle material: assemble the tool and click "YZ origin" in the V-Panel to set the tool length.

X 原點設定，origin setup：將刀具移動到工件最右端，在虛擬控制台對話方塊裡的點擊「X 原點」即可將 X 原點設定。Move the tool to the right side edge of the material and click "X origin" in the V-Panel to set the X zero.

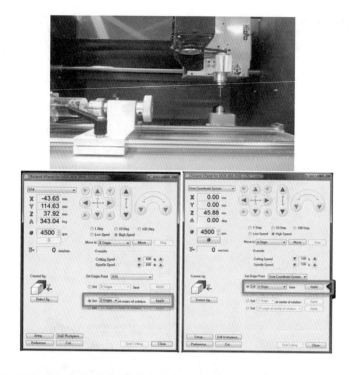

刀長及 X 座標原點設定（圓柱形材料）（Setting the Z & X axis for cylindrical block）

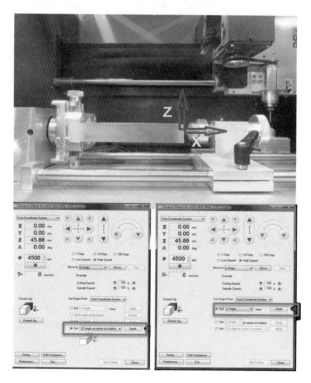

刀長及 X 座標原點設定（方形材料）（Setting the "YZ" & X axis for rectangle block）

三、輸入 NC 程式並運行（Up-load the NC program and operate）

　　基本設定完成後，在 V-Panel 對話方塊內點擊「切削」鍵，啓動「切削」對話方塊。點擊「新增」去尋找需要的 NC 程式，找到後就會出現在程式列的視窗內。重複點擊兩次所需運行的程式，在右邊的「切削」視窗內就會顯示 NC 程式的每一條指令。點擊「試車」就可以開始進行試加工。After the basic setting up, you can start machining by click "Cut" on the V-Panel. Then the "Cut" dialog box appears. To load a NC program, click "Add" bottom on the "Cut" dialog box and NC programs will appear. If you double click the NC program you are interested the G&M code will be listed in preview window. To machine, simply click "Test".

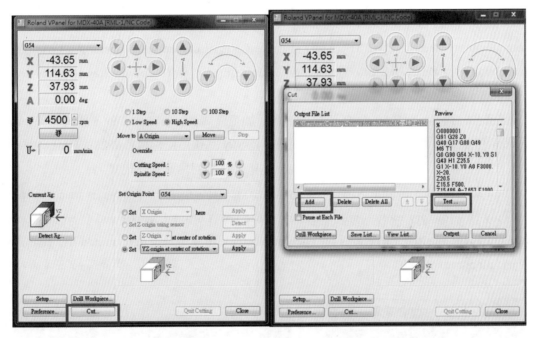

輸入 NC 程式並運行（Up-load the NC program and operate）

12.2 五軸銑床基本設定及運行實例：西門子 SINUMERIK 840D sl 控制器（Basic setup for 5-Axis Machine with Siemens SINUMERIK 840D sl Controller）

本教材以西門子 SINUMERIK 840D sl 控制器為例進行敘述。西門子 SINUMERIK 840D sl 控制器是一個高端的多軸加工機控制器，既介面友善又功能強大地實現工具機的多軸同動。下面介紹在此控制器的人機介面之下，如何方便地對加工機的工作座標以及刀具長度進行標定。High-performance milling and turning is one of the strengths of the SINUMERIK 840D sl. Further, the SINUMERIK 840D sl opens up a nearly inexhaustible technology range: From grinding and laser machining to gearwheel machining through to multitasking machining. With its superior system flexibility, the SINUMERIK 840D sl is the CNC of choice when opening up completely new technology fields. The following is the basic setup for work coordinates and tool length compensation.

此控制器目前有 X, Y, Z, A, B 和 C 軸，圖示加工機為 "Table-Table" 的 5 軸加工機，轉動方面只有沿 X 的 A 軸以及沿 Z 的 C 軸。細節可見 SINUMERIK 840D sl/828D HMI sl 銑削操作手冊。請注意這裡運動正方向的確定，均為從工件還是刀具運動為判據。

The controller can handle X, Y, Z, A, B 和 C axis. However, for the 5-axis machine in this book, only two rotatory motion are necessary, those are A around X axis and C around Z axis. **Please notify the positive directions for each movement, they are all determined by: Tool move or work-piece move**.

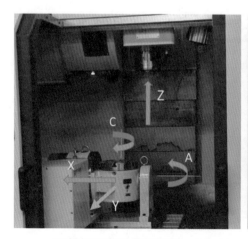 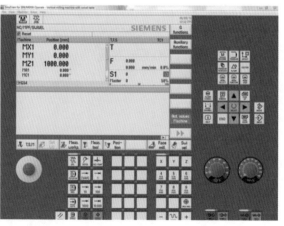

五軸加工機及機械座標，Machine Coordinate (MX1, MY1, MZ1, MA1, MC1)

*特別注意：在每次設定之前，一定要運行一次 "MDA" M Machine MDA，使得五軸轉換功能 "TRAORI" 處於非做動狀態。此外，A 軸以及 C 軸全部要歸零。

*MDA MDA 控制操作模式：Manual Data Automatic = 在自動模式中手動輸入資料 可在 MDA 模式中輸入未參照主程式或副程式之個別的程式單節或單節序列； 一按下 NC 起動按鍵，就會立即執行這些單節。

Special Attention: Before setups, run "MDA" first, M Machine MDA CYCLE START to make sure the coordinate transformation function "TRAORI" is in-active. In "MDA" mode (Manual Data Automatic mode), you can enter G-code commands block-by block and immediately execute them for setting up the machine. You can load an MDA program straight from the Program Manager into the MDA buffer. Also, A and C set to 0.

一、建立新刀具 Create new tools

作為範例，參考下圖所示建立三把刀具並設定初始長度為 0mm：量具 ,D10 以及 R3。As an example to start, create new tools Tester01, D10 and R3。Select the tools and set original lengths zero (0.0mm) as shown in the following picture.

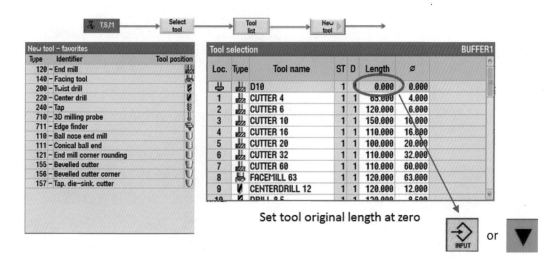

Set tool original length at zero

二、建立工件座標 Create Work Coordinate (G54---G59)

• 在控制台上點擊「選擇工件補償」，On the control panel, press "Select Work Offset" Select work offs.；

- 在顯現的工作座標表單上選擇想要設定的工作座標，On the work-offset table, select the work-coordinate you want to set（G54---G59）；

- 在控制台上點擊「手動設定」，On the control panel, click "In Manually" In Manually ；

- 再點擊 "Cycle Start" CYCLE START 按鍵完成此工作座標的建立，Click the "Cycle Start" CYCLE START key to make it happen (G54).

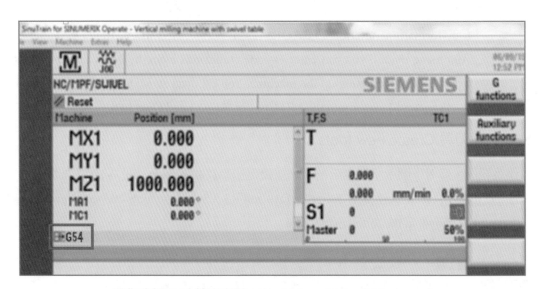

工作座標 G54 建立完畢，Work coordinate G54 created

三、設定刀具量測基準線高度 Setup gage line for Tool length (positive offsets)

接下來標定刀具長度補償值，這裡應用正補價概念。首先將主軸末端的刀具量測基準面接觸到量高器的頂端 0 點，按下 "Z=0"。打開工作座標表，將量得的 Z 座標減去 50mm，這是因爲考慮到量高器的高度 50mm¡£本例中，表內原先測定的 Z 座標是 −314.230，需要修改成 −364.230，這樣才能眞正體現刀具量測基準面接觸到工件表面的座標值。Positive tool length compansation is used here in the book. First move the tool gage line (Tool holder) to the height sensor of 50mm and click "Z=0" buttom to setup Z origin for current work-coordinate (G54---G59). Then open the Work-Offset-Table to manuely input the Z offset value, considering the height sensor thickness of 50mm. In the table, the original Z is at -314.230, manuely modify into -364.230. In this way, the gage line is surely set to the Z zero line in curent work coordinate.

主軸末端的刀具量測基準面接觸到量高器的頂端，Move the tool holder to touch the hight sensor

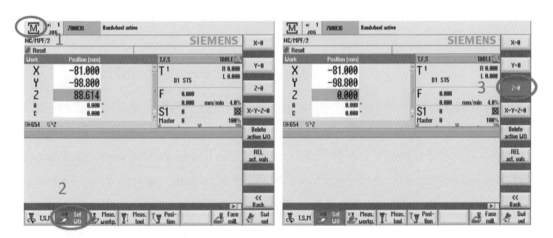

在本座標系統中設定 Z 原點，Setup Z origin for current work-coordinate (G54---G59)

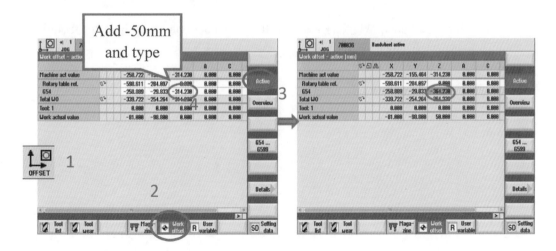

在工作座標表中手動輸入，Add the height of the sensor (50mm) into the work offset table

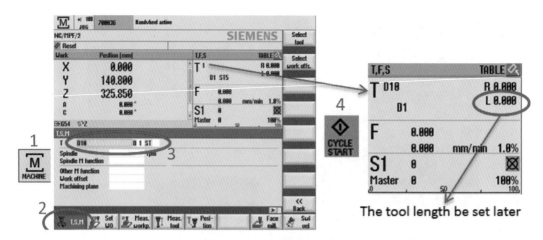

The tool length be set later

將刀具 "D10" 安裝到主軸上，可以同樣方式安裝其他刀具。Load tool "D10" onto the tool holder. Same way to load other tools and sensors.

四、設定工作座標 X,Y, Setup X, Y origin for current work-coordinate (G54---G59)

對已建立的工作座標（G54---G59）。以選擇的方式建立零點，本例採取工件中心點設定。利用任何尋邊器或百分表找到四條邊線並且分別按下 "Save P1, P2, P3, P4"，最後按下 "Save WO" 對剛剛設定的工作座標進行計算和儲存。Set X, Y origin for current work-coordinate (G54---G59). This example uses the center of a block. With any edge-finder (wiggler), find the four edges and press "Save P1, P2, P3 and P4) accordinly. Then press "Save WO" to complete the work origin setting.

注意 Note：對於用來設定 XY 座標的尋邊器來說，在相應的「刀具表單」中，其直徑「Ø」要設為 0.00mm。
For the Edge-finder (Wiggler) used in setting XY coordinates, its diameter should be set as 0.00mm i the "Tool Selection" table.

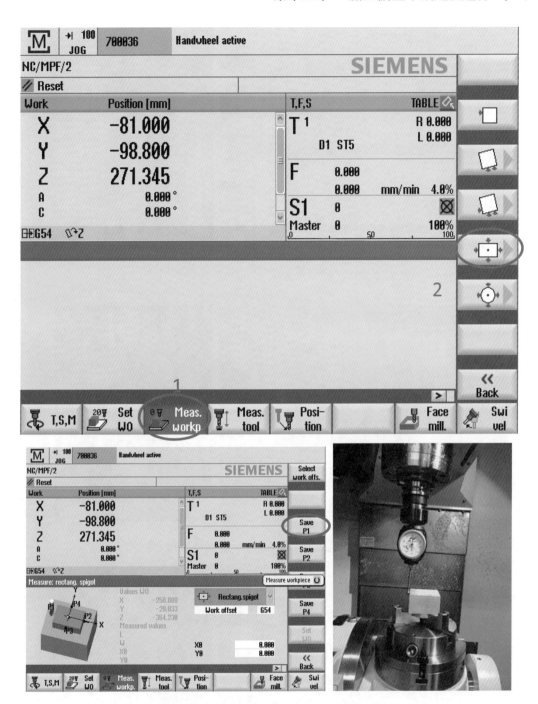

測定及設定點 P1，Sense and set Point P1

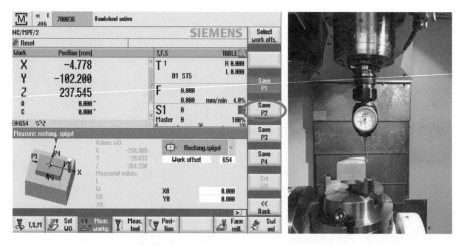

測定及設定點 P2，Sense and set Point P2

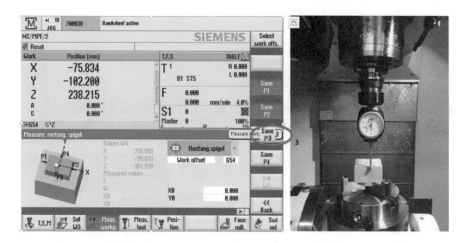

測定及設定點 P3，Sense and set Point P3

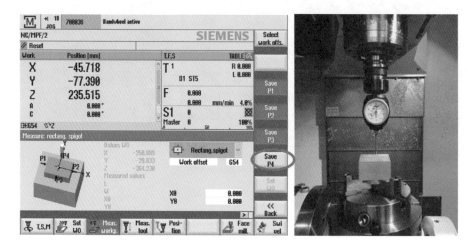

測定及設定點 P4，Sense and set Point P4

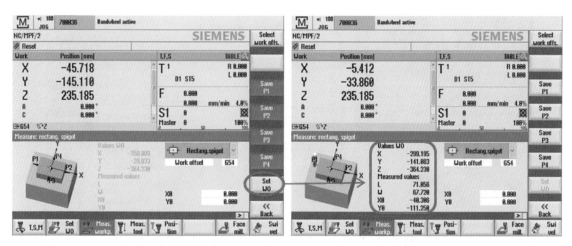

按下 "Set WO" 完成工作原點設定（Click "Set WO" to complete the setup for X, Y origin）

五、測量刀具長度（Measure the tool lengths）

刀具量測基準線高度設定後，每把刀具的長度設定就十分簡單，After the base line setup, tool length setting is quite simple:

- 先將所需標定的刀具叫入（這裡先是 D10）, Call tool "D10";
- 然後將刀尖接觸量高器並標定 0 點，Move the tool tip to the height sensor;
- 不要忘記將 "Z0" 設為 50mm，以計入量高器的高度，Do not forget to input 50 for "Z0", to offset the height sensor reading;
- 按下 "Set Length" 完成本刀具長度設定，Press "Set Length" to complete to setting;
- 需要的話，可以打開刀具表進行檢視與確認，You can view tool table to confirm;
- 重複以上三個步驟以設定刀具 R3。 Repeat above three steps for tool R3.

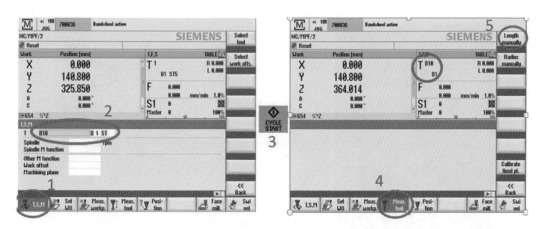

叫入刀具 "D10"（Call tool "D10"）

將刀尖接觸量高器並按「設定刀長」（Measure tool length and press "Set Length"）

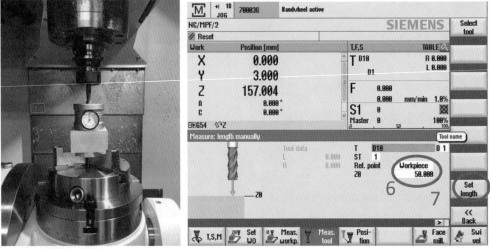

打開刀具表檢查刀長值（See length table and set radius of the tool）

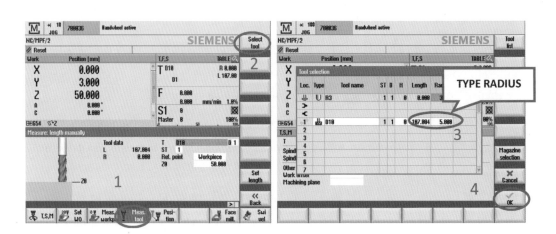

同樣方法設定刀具 R3（Repeat above steps to set tool length for toll "R3"）

六、快速移動至所設原點進行確認（Confirm the setups.）

　　爲確認以上設定的正確性，可以在機器 "Machine" 狀態下，按下 "Position" 鍵進行座標歸零動作。Quick move to current work coordinate zeros to confirm the setups.

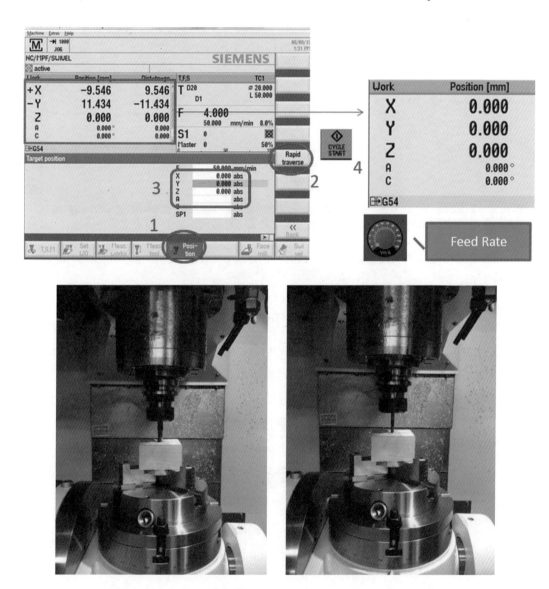

確認 D10 及 R3 刀尖歸零（Zero for tools D10 and tool R3）

七、檢視包括刀長在內的工作座標所有設定值，View current work coordinate values for specific tools.

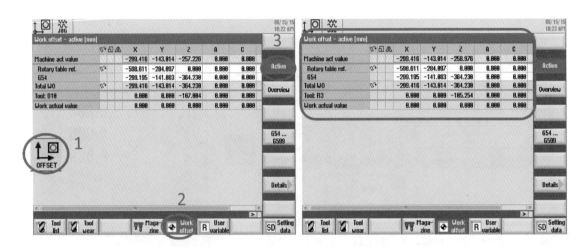

檢視工作座標所有設定值（View current work coordinate values for tools）

八、輸入 NC 程式並運行（Import NC program and run）

機器設定好以後就可以根據 NC 程式開始加工。在控制台上，On the control panel:

• 按「程式管理」鍵，Press "Program Manager" button;

• 選擇所需的 NC 程式，Select the NC program you want to run;

• 按「開啟」鍵打開程式，Press "Open" and the NC program appear on the screen;

• 確認無誤後按下「執行」，Then press "Execute" to activate the program;

• 必要時也可以在螢幕上先進行模擬，You can also simulate on the screen;

* 建議剛開始時使用「單節運行」的模式，等到一切確認以後再進行「連續運行」。

During starting period, the "Single Line" mode is always suggested for safety reason. Conduct continues running mode when everything is certain with confidence.

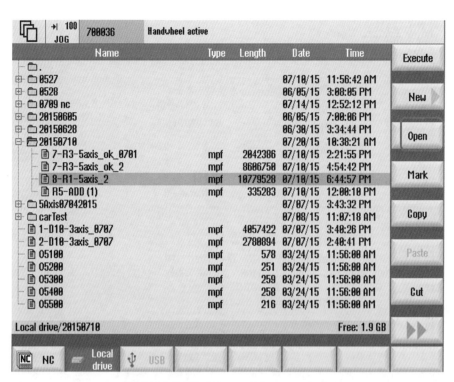

開啟 NC 程式（Open NC program）

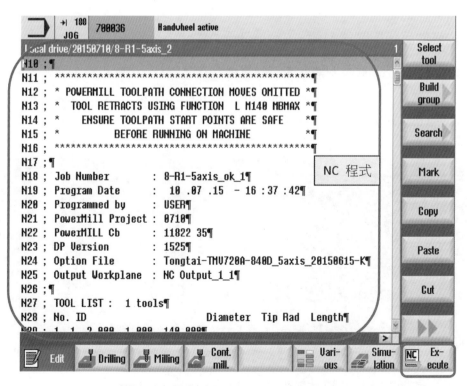

執行 NC 程式（Execute NC program）

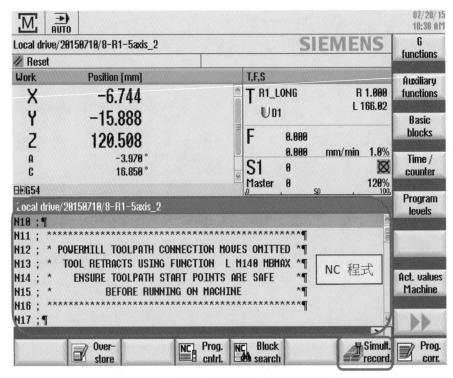

模擬 NC 程式（Simulate NC program）

九、一些其他品牌控制器範例，Controllers of other brand

　　以下是一些其他品牌控制器的面板的範例，設定基本上大同小異，需要確定的是三軸以上的多軸控制器。There are many controllers of other brands for CNC machining, the work coordinate and the tool length setup are basically similar.

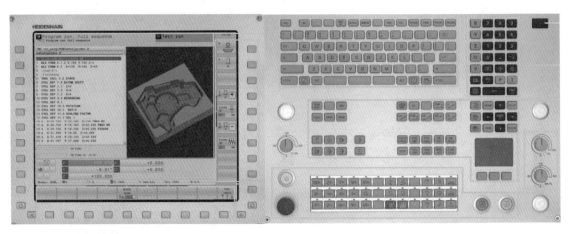

海德漢控制器面板，德國（HEIDENHAIN Control Panel, Germany）

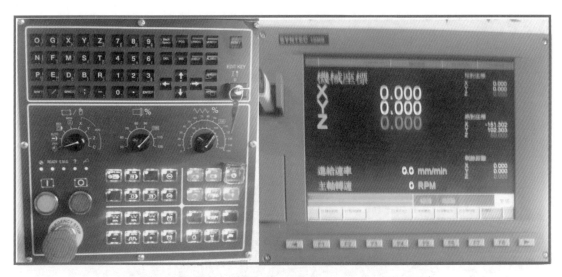

SYNTEC 控制台，臺灣（SYSTEC Controller, Taiwan）

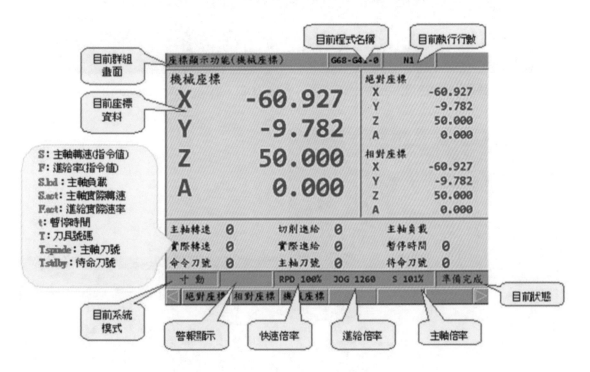

台達電四軸加工控制台操作介面，臺灣（DELTA Controller, Taiwan）

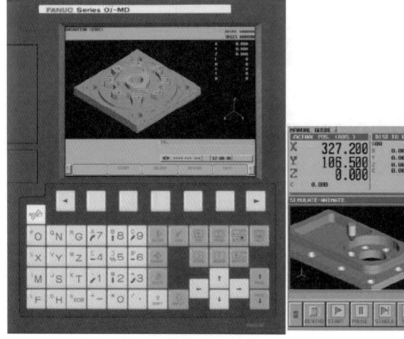

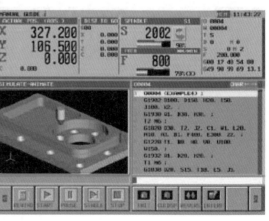

發那科控制器，日本（FANUC Controller, Japan）

參考文獻（References）

1. Roland MDX-40A CNC 雕刻機銑削操作手冊

2. SINUMERIK 840D sl/828D HMI sl, Milling，西門子銑削操作手冊

3. 840D sl SINUMERIK Operate V2.7/4.4/4.5, 5-axis transformation TRAORI，西門子銑削操作，
   五軸轉換（TRAORI）

4. 840D sl SINUMERIK Operate V2.7/4.4/4.5:
   Advanced Surface (CYCLE832)，西門子銑削操作，先進表面加工（CYCLE832）

5. http://www.heidenhain.tw/

6. http://www.delta.com.tw/product/em/em_press_detail.asp?nid=914

7. http://www.syntecclub.com.tw/2010/HTML/Product.aspx

8. http://www.fanuc.eu/uk/en

附錄-1　G&M碼基本機能簡介
（List of Basic G&M Codes for Milling）

機能指令，並歸類為六大類，There are six categories of codes:

- G 機能（準備機能：G00～G99），G Code;
- M 機能（輔助機能：M00～M99），M Code;
- T 機能（刀具機能），T Code;
- S 機能（主軸轉速機能 or 切削機能），S Code;
- F 機能（進給機能），F Code;
- N 機能（程式序號機能），N Code.

常用 G 碼一覽表，List of G-codes commonly found in CNC controls

Code	描述	Description
G00	快速定位	Rapid positioning
G01	直線切削	Linear interpolation
G02	圓弧切削／螺旋切削（順時針）	Circular interpolation, clockwise
G03	圓弧切削／螺旋切削（逆時針）	Circular interpolation, counterclockwise
G04	暫停	Dwell time
G05	高速高精度加工指令	High-precision contour control (HPCC)
G08	先行控制	Spline Smoothing On
G09	真確停止	Exact stop check, non-modal
G10	可程式資料輸入	Programmable data input
G11	可程式資料輸入取消	Data write cancel
G15	極座標 OFF	Polar coordinate system, OFF
G16	極座標 ON	Polar coordinate system, ON
G17	XP - YP 平面選擇	XY plane selection
G18	ZP - XP 平面選擇	ZX plane selection
G19	YP - ZP 平面選擇	YZ plane selection
G20	英制單位輸入	Programming in inches
G21	公制單位輸入	Programming in millimeters (mm)
G28	自動原點復歸	Return to home position (machine zero)

Code	描述	Description
G30	第二、三、四參考點位置回復	Return to secondary home position (machine zero)
G31	跳躍機能	Skip function (used for probes and tool length measurement systems)
G33	螺紋切削	Constant-pitch threading
G40	刀具半徑補正取消	Tool radius compensation off
G41	刀具半徑補正（左補償）	Tool compensation to the left
G42	刀具半徑補正（右補償）	Tool compensation to the right
G43	刀具長度補正（負方向）	Tool length compensation - negative direction
G44	刀具長度補正（正方向）	Tool length compensation - positive direction
G49	刀具長度補正取消	Tool length compensation cancel
G50	比例放大、縮小 ON	Define the maximum spindle speed
G51	比例放大、縮小 OFF	Part rotation; programming in degrees
G52	局部座標系設定	Local coordinate system (LCS)
G53	機械座標系選擇	Machine coordinate system
G54-G59	第一至六工件座標系選擇	Work coordinate systems (WCSs)
G61	真確停止模式	Exact stop check, modal
G62	自動轉角加減速	Spline contouring with buffering mode off
G64	切削模式	Default cutting mode (cancel exact stop check mode)
G68	座標系統旋轉開啟	Rotation ON, 2D/3D
G69	座標系統旋轉取消	Rotation OFF, 2D/3D
G73	高速分段鑽孔循環	Fixed cycle, multiple repetitive cycle, for roughing, with pattern repetition
G74	左螺旋攻牙循環	Peck drilling cycle for turning
G76	精搪孔循環	Fine boring cycle for milling
G80	固定循環取消	Cancel canned cycle
G81	點鑽孔循環	Simple drilling cycle
G82	沉頭孔循環	Drilling cycle with dwell
G83	分段鑽孔循環	Peck drilling cycle (full retraction from pecks)
G84	右螺旋攻牙循環	Tapping cycle, right hand thread,M03 spindle direction
G85-	搪孔循環一至四	Boring cycle 1 to 4
G89	盲孔鉸孔循環	Boring with intermediate stop canned cycle
G90	絕對值座標系統	Absolute mode
G91	增量值座標系統	Incremental mode
G92	工件座標系建立與更變	Home coordinate reset

Code	描述	Description
G98	固定循環起始點復歸	Return to initial Z level in canned cycle
G99	固定循環 R 點復歸	Return to R level in canned cycle

M 機能一覽表（List of M-codes commonly found in CNC controls）

Code	描述	Description
M00	程式停止	Compulsory stop
M01	選擇性停止	Optional stop
M02	程式結束	End of program
M03	主軸正轉	Spindle on (clockwise rotation)
M04	主軸反轉	Spindle on (counterclockwise rotation)
M05	主軸停止	Spindle stop
M06	刀具交換	Automatic tool change (ATC)
M07	油霧切削開	Coolant on (mist)
M08	切削液開	Coolant on (flood)
M09	油霧切削／切削液／主軸中心出油／油路刀桿關	Coolant off
M10	第四軸鎖緊	Pallet clamp on
M11	第四軸放鬆	Pallet clamp off
M12	主軸中心出油／油路刀桿開	Milling spindle mode cancel (Turning mode selection)
M13	主軸正轉、切削液 ON	Spindle on (clockwise rotation) and coolant on (flood)
M14	主軸正轉、切削液 ON	Milling tool reverse rotation
M15	刀庫刀套上升	Milling tool stop
M16	刀庫刀套下降	Spindle orientation 0° (for AJC)
M17	刀長量測開／工件量測關	Spindle orientation 120° (for AJC)
M18	刀長量測關／工件量測開	Spindle orientation 240° (for AJC)
M19	主軸定位	Spindle orientation
M20	主軸定位解除	Start oscillation (configured by G35)
M21	備用	Mirror, X-axis/ Tailstock forward
M22	備用	Mirror, Y-axis/ Tailstock backward
M23	備用	Mirror OFF/ Thread gradual pullout ON
M29	剛性攻牙	Rigid Tapping Mode on Fanuc Controls
M30	程式結束及回復	End of program, with return to program top

Code	描述	Description
M31	自動進給上限資料傳遞設定（製造商專用）	Tail spindle & tailstock body advance (for 300/400-III/IIIT)
M32	自動進給控制 AFC 關	Tail spindle & tailstock body retract (for 300/400-III/IIIT)
M33	自動進給下限資料傳遞設定（製造商專用）	Low chuck pressure
M34	刀具號碼重整（製造商專用）	High chuck pressure
M40	螺旋排屑 ON	Automatic spindle gear range selection
M41	螺旋排屑 OFF	Spindle gear transmission step 1
M48	有效切削進給率調整	Feedrate override allowed
M49	無效切削進給率調整	Feedrate override NOT allowed
M52	夾治具夾（客戶選用功能）	Unload Last tool from spindle
M60	工件交換台交換	Automatic pallet change (APC)
M70	DNC ON	Spline definition, beginning and end curve 0
M71	DNC OFF	Spline definition, beginning tangential, end curve 0
M80	刀具號碼設定準備	Delete rest of distance using probe function, from axis measuring input
M81	刀具號碼重新設定	Drive On application block (resynchronize axis position via PLC signal during the block)
M98	副程式呼叫	Subprogram call
M99	副程式結束	Subprogram end

參考文獻（References）

1. https://en.wikipedia.org/wiki/G-code
2. http://www.helmancnc.com/mazak-integrex-m-code-list/

附錄-2　多軸銑削加工認證試題
（Certification Tests for multi-axis machining）

多軸銑削加工認證試題(一) － 手錶
Certification for multi-axis machining Test (1)-A Watch

加工條件 Machining Conditions:

1. 加工材質，Material: NAK55 (HRC40);
2. 刀具廠商：自定自選，Tool Supplier: Any;
3. 物件基準；底面中心，Datum: Bottom center;
4. 成品光潔度：Surface finsh: Ra 0.2~0.4μm;
5. 最小使用刀具，Smallest tool: D1

評比標準，Criteria:

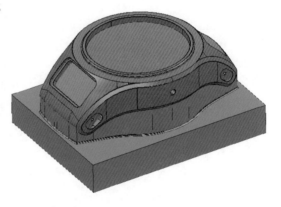

項目，Items		評核分數，Score
路徑策略 Tool path Strategy	粗加工，Rough	5
	再次粗加工，Rough-2	10
	中加工，Semi	10
	細加工，Finish	10
	清角加工，Corner Cleaning	10
	逃料加工	10
路徑品質，Tool path quality		20
加工安全，Safety		10
預估時間，Time		10
其他評核，Others		5
總計		100

多軸銑削加工認證試題(二)－多面體
Certification for multi-axis machining Test (2)-Polyhedron

加工條件：

1. 加工材質，Material: NAK55 (HRC40);

2. 刀具廠商，自定自選，Tool Supplier: Any;

3. 物件基準，底面中心 Datum: Bottom center;

4. 成品光潔度，Surface finsh: Ra 0.2~0.4μm;

5. 最小使用刀具，Smallest tool, Ball nosed tool: R1

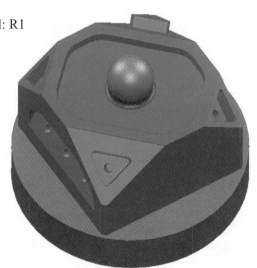

評比標準，Criteria:

項目，Items		評核分數，Score
路徑策略 Tool path Stratagy	粗加工，Rough	5
	再次粗加工，Rough-2	10
	中加工，Semi	10
	細加工，Finish	10
	清角加工， Corner Cleaning	10
	逃料加工	10
路徑品質，Tool path quality		20
加工安全，Safety		10
預估時間，Time		10
其他評核，Others		5
總計		100

多軸銑削加工認證試題(三) − 多孔鑽
Certification for multi-axis machining Test (3)-Multiple Drilling

加工條件：

1. 加工材質，Material: NAK55 (HRC40);
2. 刀具廠商，自定自選，Tool Supplier: Any;
3. 物件基準，底面中心，Datum: Bottom center;
4. 成品光潔度，Surface finsh: Ra 0.2~0.4μm;
5. 最小使用刀具，Smallest tool, Ball nosed tool: R1

評比標準，Criteria:

項目，Items		評核分數，Score
路徑策略 Tool path Strategy	粗加工，Rough	5
	再次粗加工，Rough-2	10
	中加工，Semi	10
	細加工，Finish	10
	清角加工， Corner Cleaning	10
	逃料加工	10
路徑品質，Tool path quality		20
加工安全，Safety		10
預估時間，Time		10
其他評核，Others		5
總計		100

國家圖書館出版品預行編目資料

五軸銑削數控加工之基礎及實作／王松浩,
吳世雄著. -- 二版. -- 臺北市：五南,
2019.04
　　面；　公分
　　ISBN 978-957-763-337-8 (平裝)

1.機械製造　2.電腦程式　3.電腦輔助設計

446.8927029　　　　　　　108003517

5F64

五軸銑削數控加工之基礎及實作

作　　者 — 王松浩（6.5）　吳世雄

發 行 人 — 楊榮川

總 經 理 — 楊士清

主　　編 — 王正華

責任編輯 — 金明芬

封面設計 — 簡愷立　王麗娟

出 版 者 — 五南圖書出版股份有限公司

地　　址：106台北市大安區和平東路二段339號4樓

電　　話：(02)2705-5066　　傳　　真：(02)2706-6100

網　　址：http://www.wunan.com.tw

電子郵件：wunan@wunan.com.tw

劃撥帳號：01068953

戶　　名：五南圖書出版股份有限公司

法律顧問　林勝安律師事務所　林勝安律師

出版日期　2015年11月初版一刷
　　　　　2019年 4 月二版一刷

定　　價　新臺幣400元